The Non-Toxic Farming Handbook

Philip A. Wheeler, Ph.D.
Ronald B. Ward

The Non-Toxic
Farming Handbook

Philip A. Wheeler, Ph.D.
Ronald B. Ward

Acres U.S.A.
Austin, Texas

The Non-Toxic
Farming Handbook

Acres U.S.A.
P.O. Box 91299
Austin, Texas 78709 U.S.A.
(512) 892-4400 • fax (512) 892-4448
info@acresusa.com • www.acresusa.com

Printed in the United States of America

Publisher's Cataloging-in-Publication

Wheeler, Philip A. 1939-
    The Non-Toxic Farming Handbook / Philip A. Wheeler &
Ronald B. Ward. — 2nd ed.
        p. cm.
        Includes bibliographical references and index.
        ISBN: 0-911311-56-4

    1. Organic farming. 2. Sustainable agriculture. 3.
Organic fertilizers. I. Ward, Ronald B., 1944- II.
Title.

S605.5.W44 1998   631.5'84
                                        QBI98-1051

Library of Congress Catalog Card Number: 98-73502

Dedication

This book is dedicated to all the enlightened pioneers: scientists, consultants, growers, environmentalists, teachers, philosophers and others who preceded us. Without their insights, discoveries, experiences and dreams we would not have been able to achieve our still limited understanding of how food, fiber and livestock can be raised to the benefit of mankind without causing damage to ourselves and our soil, water and air.

— the authors

The Non-Toxic Farming Handbook

Table of Contents

Acknowledgements ix

Foreword. xi

The First Cause xvii

Introduction 1

Chapter 1   What the Experts Say 5

Chapter 2   Energy 11

Chapter 3   The Soil Environment 15

Chapter 4   Weeds. 21

Chapter 5   Insects. 27

Chapter 6   Soil Testing 33

Chapter 7   Interpreting Soil Test Information. 43

Chapter 8   Making Fertility Recommendations 59

Chapter 9   What's Wrong with Today's Fertilizers? 65

Chapter 10   Blended Fertilizers 71

Chapter 11   Selecting the Correct Fertilizer 77

Chapter 12   pH, Lime & Calcium 101

Chapter 13   Moon Cycles & Plant Manipulations 109

Chapter 14   Refractometer 117

Chapter 15   Foliar Feeding 123

Chapter 16   The Role of Nutrition 131

Chapter 17   Water 137

Chapter 18   Forage 145

Chapter 19   Livestock Nutrition 149

Chapter 20   Tillage 169

Chapter 21   Field Evaluation 181

Chapter 22   How to Recover a Field 187

Chapter 23   Levels of Non-Toxic Farming 195

Chapter 24   Suggested Fertility Programming 201

Chapter 25   Subtle Energies 213

Resources 221

Bibliography 223

Index 227

Acknowledgements

The authors would like to acknowledge and thank the contributors to these specific topics: Jerry Brunetti — livestock nutrition; Jay McCaman — weeds and tillage; and Tom Besecker — tillage.

We also acknowledge all the growers, distributors and dealers in the U.S. and other countries who worked with us during the TransNational Agronomy experience, whose valuable feedback helped us to evaluate our ideas of how nature works.

And finally we would like to acknowledge and thank Richard (Dick) Vaughan, the third member of the TransNational Agronomy team, who allowed us to work on this book while he kept the business wheels turning.

Foreword

The late Albert Schweitzer once told an audience that demonstration was not only the best way to teach a subject, it was the only way. The writers of *The Non-Toxic Farming Handbook* must have had such a dictum in mind. Their 25 chapters are at once a manual for hands-on agronomy and a blackboard lecture backgrounding the reason for being of non-toxic farming.

Some of the principles discussed by Wheeler and Ward trace back to the beginning of the present century. Others reach into a tomorrow still to be discovered by academia.

In this superb volume, these scientists and consultants have examined and described two basic premises upon which toxic agriculture was constructed in the first place, namely the mandate to annihilate weeds and insects alike. They point out that every attempt to destroy the creatures that rent space on the farms or in the human body is a two-edged sword. Destruction of friendly organisms is a terrible price to pay for termination of the harmful ones. The well individual — corn crop or human being — is the one with less harmful renters than beneficial ones.

We can trace the biological equation with stark brevity. For it is the health of the human being that makes the final connection to the soil.

Among the beneficial creatures attending the human being are the red corpuscles. Born in the marrow of bones and in the spleen, the little reds live about 30 to 90 days. They come and go the way human beings live and die. They collect oxygen from the air by going to the lungs for that resource and then they deliver it to all parts of the body. Nourishment on tap by cells needs oxygen to form.

White blood cells are veritable partners with the little reds. It is their function to hunt and destroy. The enemy is any harmful pathogen. From a white blood cell's point of view, we live in a cesspool. The air we breathe is loaded. The water is polluted. Pathogens reside on our skin. They even hide between other molecules. Should these invaders find a weak spot, they set up shop and proliferate. White corpuscles do not take these invasions lightly. When the absconders arrive, white blood cells tear them to pieces. They can't lose if the body keeps on hand plenty of arms and ammunition. If, however, nutrients are imbalanced or deficient, the white defenders fight a losing battle and disease follows.

This is the basic story of mankind. What Wheeler and Ward have done in this handbook is present the same story, now enlarged to include agriculture, crops, animal husbandry, energy exchange, and the anatomy of weed and insect control, without poisons, of course. In so doing, they have bolted the head of the biotic pyramid to the soil and to the nutrition man takes from the soil.

The leap is long overdue. Ever since toxic agriculture came to command the nation's universities, great strides in learning have been canceled out. Professors no longer remember, if they ever knew, that "phosphorus for the nerves, calcium for the bones" were copybook maxims. It is no surprise that too few farmers comprehend the difference between organic and inorganic minerals. Calcium may be fine for a whitewashed fence, but it isn't food. First it has to be changed into organic calcium either via the biology of a plant or the microorganism of a gut. A shingle nail won't provide iron, and organic iron won't melt down to cast into a nail. "Animals are not mineral eaters," the late William Albrecht used to say. "They head for the mineral block only as an act of desperation."

One of the giants who stood at the crossroads of change was Carey A. Reams. His matchless science backgrounds many of the chapters in this handbook. Certainly his use of the refractometer to assess health and crop quality ranks as a premier development in the practice of non-toxic agriculture. Phil Callahan's contributions to the paramagnetic-diamagnetic discussion is also backbone material in the pages that follow, as are the energy equations so-called conventional agriculture has still to understand. But these are hands-on chapters and leave unstated the reasons for our continuing uphill battle. With all the increased knowledge available, we are required to wonder why change remains an arm's length away.

Perhaps the late C.J. Fenzau, writing in an early issue of *Acres U.S.A.*, best plumbed the depths of the natural psyche. "When living life energies unite as an ultimate objective for our nation, only then can we begin to construct the understanding of the ecology of the soil and be able to benefit from its minuteness and wondrous patience. Nature has no limit on time. She is patient and forgiving. She is able to repair herself from the ignoble treatment of man in spite of his tremendous physical capacity for destruction. As we continue to replace nature, we assuredly prevent the development of our mental capacity to learn and fully complement nature, a requirement expected from us in permission for life."

The above is a quote from the first year of *Acres U.S.A.*'s publication. Since then the practice and science of eco-agriculture has been refined to the point where this handbook can be considered a crowning achievement. And yet through it all, certain basics remain in place, even after being enlarged and validated by Wheeler and Ward.

A soil system for energy and nutrient production is a living system in which bacteria and other soil organisms must receive nutrients and energy from proteins, carbohydrates, cellulose, lignins, all organic materials from a soil that has a managed supply of air and water within a balanced chemical environment. This chemical balance involves more than simplistic N, P and K. It requires an equilibrium of pH, calcium, magnesium, sodium and potassium, humus and a nutritional balance of sulfur, with correct relationships of nitrogen to calcium, calcium to magnesium, etc. Many readers will recognize the above as an Albrecht conceptualization.

Most of this handbook has to do with the Reams system for calculating energy and titrating microscopic bodies as computed to crop acres. The system is dazzling in its purity and has proved itself to those who are willing to pay out the price of an effort.

And the great professor Phil Callahan says, "No method of insect control will ever work as long as poisoned crops outgas ethanol and ammonia in small parts per million. Those two powerful fermentation chemicals are the mark of a dying, decaying plant and serve as attractants to all plant-eating insects."

"Ultimately, a healthy crop is the best insurance against disease and insect damage." Callahan continues, "The disappearance of high cgs readings has assured that very little of the good fertilizers and minerals dumped on farmland is utilized. That is why our rivers are polluted from unutilized runoff."

Perception and reality have merged in eco-farming during recent years, and this is the reason for being of this handbook. When the lessons are applied to energy management, weed and insect control, fertility computations and the grab bag called crop production, they shout aloud, "We have come of age."

Withal, the leaps in scientific knowledge made a matter of record in this handbook do more than answer a reality that has bedeviled eco-farming from its inception. When *Acres U.S.A.* first went to press, there were few trained soil scientists who understood or could extend a comprehensive judgement — other than lip service — to the standard viewpoints important to human health, namely soil bacteria, earthworms, soil drainage, tilth, soil structure, and the true function of organic material in the soil system. They could state, however, that organic matter converts, regulates or releases nutrients to soil life which in turn makes available biological nutrients to the bio-system of the soil from which man was originally evolved. For many years, organiculture was downgraded because it had only amateurs who taught others to be good amateurs.

This defeat has now been rectified. There isn't an agronomist in the university more capable than the authors of this handbook. We'll let it rest with that statement. The backup details are now a matter of record.

— *Charles Walters, founder, executive editor,* Acres U.S.A.

The First Cause

The authors share a deep personal belief that the ultimate cause behind all that is revealed to us is commonly referred to as God. As Spirit, God reveals Godself in multiple ways, not the least of which is nature.

God (nature) has set immutable laws which govern man and his actions. To understand these laws and to work within their confines is to receive the blessings inherent within the laws themselves. To ignore them, regardless of motivation, is to suffer the consequences.

The authors do not claim to fully understand the laws of nature. We do, however, feel blessed in having had opportunities to work with some who understand more than we.

You can consider the Divine Source as a form of cosmic subtle energy or not. God's presence is subtle energy; God is much more.

May God grant His blessings to all who read and understand this book.

Introduction

Today's farmer has the knowledge and technology to farm differently than his father. Having passed through promises of high yields through chemistry, he looks back on the fulfilled and unfulfilled prospects of the "modern" farming era. It is true that modern farm chemicals have had a real effect on farming. Farming has changed dramatically through their use — yields have increased, weed pressure has supposedly been controlled, and insect, nematode, and fungus damage have reportedly been held at bay — all enabling larger farms to be worked by fewer individuals. However, are these claims true and has all of this been good? Are we really better off today? What has been the trade-off?

To answer this last question will require us to wait, perhaps several decades, while we uncover and tally the damage done to man and the environment caused by the use, misuse, and overuse of today's toxic chemicals. Even non-pesticide-type materials, such as chemical fertilizers, have caused damage as countless wells have been declared unsafe for human use due to the high concentration of nitrates. Great damage has been caused by the concentrated toxic chemicals (previously advertised as "safe") and the use of the seemingly safe and necessary fertilizers, especially anhydrous ammonia, 0-46-0 and 0-0-60.

Today's farmers have increased rates of selective cancers and other degenerative diseases. In-depth training is now required in the proper and safe application of chemicals. The Environmental Protection Agency (EPA) has been slow to fully test the many chemicals in the marketplace. The EPA has also failed to test combinations of chemicals applied separately but frequently enough to the same fields that active ingredients unite together forming unknown combinations in the soil.

You will find some basic information and down-to-earth comments, suggestions, and observations in these pages. Much of this material comes from Dr. Carey Reams, a brilliant consultant and researcher who made astounding findings in human and animal nutrition and soil and crop fertility. His findings regarding the concept of energy, the important nutrient roles of calcium and phosphorus, and the measurement of plant nutrition via the use of the refractometer have altered and will continue to alter the course of agriculture.

Credit must also be given to Dr. Philip Callahan, retired USDA entomologist and researcher based at the University of Florida, Gainesville. His contributions have given us a better understanding of how insects are attracted to plants. Whereas, his work on infrared radiation has contributed to the development of military technology; it is now time to apply his work to the field of agriculture where he intended it in the first place.

Additional thanks go to Dr. T. Galen Hieronymus for his pioneering work on an instrument that measured eloptic energy. This device is known as a radionics instrument or, as we refer to it, an electronic scanner. Patents have been obtained from the United States, Canada, and even England on this invention.

The information presented herein is for use by any agricultural system that wishes to grow crops with fewer or no toxic chemical inputs and seeks to enhance the feed and food value of the crops grown. The methods described are not meant to be a definitive treatise on how to farm, but are meant to be good, sound suggestions that can and have been successfully used by hundreds of farmers. Due to the variety of soil, weather, and management conditions found on each farm, no claim is made that using any or all parts of this system will result in a superior crop in quality or quantity.

No inference is made that there are not other agronomic systems that may be equal or superior in value for a given situation. It should also be noted that the discussion herein is considered an alternative method to standard agronomic practices as currently taught and practiced and should be approached with the caution due any new or different idea.

We are aware that women play a vital role in farm management. The use of the pronoun *he* has been selected for convenience only. Please don't construe this usage as a slight or unawareness.

Chapter 1

What the Experts Say

There are basically two sources of so-called "expert" information available concerning agriculture. These consist of private enterprise and the public sector which includes the USDA and other governmental agencies, land grant universities, and the agricultural extension service. Private enterprise is self-serving, so their recommendations need to be scrutinized. Can you imagine, for instance, Dow or Monsanto advocating non-toxic farming? One only has to read current publications from universities and extension services to discover what their best judgments are regarding the use of farm chemicals.

Information Sources

Commercial companies and governmental authorities have given basically the same message: farming is economically infeasible without the use of modern chemicals including the acid and salt fertilizers and the toxic herbicides, pesticides, and fungicides. Chemical companies pay significant sums for university experimentation to test their products prior to marketing. Consequently, the university professor and extension agent often have the latest information on the chemicals in question.

People like Rachel Carson, author of *Silent Spring*, and others became extremely critical regarding the long-term consequences of public policies which emphasized, to the exclusion of other approaches, agricultural programs dependent upon toxic farm chemicals during the past two or three decades. Is there any truth behind their conclusions?

During the end of the Carter administration, Dr. Garth Youngberg was hired to head a department within the USDA to research and advise on issues related to organic farming. His federal employment was short-lived as the Reagan administration disbanded the position. It was obvious that the politics of the era over-rode the growing awareness of a need for change.

Recent history has seen a new and closer look being taken toward the use of "safe" farm chemicals. Recent disclosures regarding grain sprayed with a fumigant, which subsequently caused serious health problems, prompted national manufacture recalls removing cake mixes and breads from grocery shelves. The Gerber baby food company announced it would not buy apples sprayed with certain chemicals. Even herbicide advertisements are changing their message. A recent advertisement claims "less stress" than competitors, whereas they previously claimed no negative effects on crops. Some farmers were used to mixing spray solutions with their bare hands, and they now are being admonished to "read the label," use extreme caution, and properly dispose of containers.

Media attention has been heavy in the coverage of such issues as the Love Canal contamination. EPA statistics indicate that fully 10 percent of rural wells are contaminated with farm chemicals. Ground water contamination means long-term consequences of the highest magnitude. Recently, the public has been frightened over the use of Alar, a chemical used to grow apples. The use of this chemical has now been discontinued. Even the National Academy of Sciences has declared that farming must change and move to a less toxic, more sustainable approach. It is publicity like this which is rapidly changing the politics of farming.

The New Trend

The government's response was to move in the direction of LISA. LISA, short for Low-Input, Sustainable Agriculture, was an attempt to respond to public pressures for policy change. Several programs have come and gone since LISA. Now that the public's voice is being heard equally as strong as the voice from big business, politicians are pressing universities to explore other options. The university response has been to write increasingly more about non-toxic approaches to farming including giving prominence to such "alternative agriculture" priorities as soil humus and microbial life.

The momentum of giant national and multi-national manufacturing companies assures the continued sale and use of toxic materials for a long time to come. Many of the chemical companies are now buying seed companies so they can breed plants which tolerate their chemicals. These kinds of activities need to be exposed as well as offering information on how to eliminate the need for using these toxins in the first place.

Conventional thinkers decry the ability of a vocal celebrity, such as was used in the Alar controversy, to substantially alter the buying patterns of society. They point to evidence showing how safe the chemical is. Yet public opinion is changing. The fact that science cannot accurately trace a substance and clearly show resulting damage, as in the use of tobacco, is no longer being considered just cause for continuing a practice. Even as this book goes to print, scientists admit that radiation (from many sources, such as power lines, airports, computer monitors, and microwave ovens) is harmful at much lower levels than was previously acknowledged. The Russians have known this for years and have set their standards for radiation considerably lower than we have in the United States.

The public has lost a great deal of its trust in science and government agencies and is calling them to task. The people side of science and technology issues are now being discussed. What do we consider a holistic fertility approach? Are any synthetically-derived chemicals good? Are we talking about organic approaches? In the remainder of this handbook, we will address these and similar questions. We will provide information which shows that some non-toxic manufactured materials are extremely beneficial, some "natural" or mined substances are harmful, and that many of the toxic chemicals simply are not needed in the first place. The use of toxic rescue chemistry has gone full circle and now is heading toward its own end. Technology exists today which makes the need for these toxic chemicals almost a thing of the past.

The New Experts

Although the USDA and land grant universities are putting considerable energy into "catching up" with current public concerns, they will continue to be viewed with suspicion for a long time. The old adage "a rose by any other name" is difficult to escape. Yet there is a new breed of experts who can be consulted. These people, probably many of whom are unknown to the authors, have been working all along but haven't drawn the attention of the larger public.

One of the larger forums to access and utilize this new technology is *Acres U.S.A.* A monthly magazine promoting eco-agriculture, *Acres U.S.A.* is actually an institution in itself. Driven by the insights of Charles Walters, this institution for the betterment of farming has promoted non-toxic farming for over a quarter century. Subscriptions to this monthly publication can be obtained by contacting the office listed in the bibliography.

At a recent *Acres U.S.A.* conference, one of the speakers, Dr. Arden B. Andersen, discussed what he considered to be the difference between conventional agriculture and biological agriculture, or as he termed it, "nature." He stated that conventional agriculture views nature as a linear system or the whole is equal to the sum of its parts. Scientists take one aspect, the nitrogen needs of plants for instance, and believe they can research only that one variable. Once they discover an answer, they then know what to do — in this case, how much nitrogen to apply to a given crop. Thus, we have the different agronomic department specialties, such as soil science, entomology, horticulture, etc., all existing within the university structure with each researching its own questions.

Seldom do these departments communicate together because such interaction would violate the "turf" of another department.

"Nature is actually non-linear, or the whole is greater than the sum of its parts," claims Dr. Andersen. To view one aspect of nature requires the observer to consider all other variables. Only then can conclusions be made. If this approach were used, the authors believe, many of the simplistic conclusions given by the "experts" would have been "corrected" long ago.

Support for this view also comes from Deane Juhan, author of *A Handbook for Bodywork*. He writes, "Fixed laws and fixed forms are reassuring certainties only as long as they are not brought into too close a contact with the actualities of physical process. When they are, peculiar and unsettling things happen, because nature refuses to be either simple or precisely repetitious. For instance, if we are to grant on the one hand that all physical forms are subject to a universal and continuous network of causes and effects, then we are also forced on the other hand to admit that no single event has ever been repeated exactly the same way twice, because the surrounding conditions have always shifted. This in turn has a curious effect upon the notion of reliably repeatable experimental results; many real variables and many real effects must be left out for the sake of consistency. No matter how analytically useful strict logic and precise measure may be, they have very little to do with the furious interactions and the delicate dynamic stabilities of matter."

Much of the information presented in this handbook comes from the teachings of Dr. Carey A. Reams. Dr. Reams was born in Orlando, Florida in 1904. He spent most of his lifetime as a researcher, agronomist and student of human and animal nutrition, eventually overcoming great obstacles to develop his Biological Theory of Ionization.

While serving as a chemical engineer in the Philippines during WWII, the jeep in which he was riding ran over a land mine. He awoke 31 days later on the operating table having been crushed from the waist through the pelvis. His right eye was gone; he had lost all his teeth; his jawbone was fractured; his neck was broken; his back was broken in two places; and the lower part of his body was completely paralyzed. Physicians told him he would never walk again.

In December, 1950, after more than 40 surgeries, Dr. Reams was told by his physician to return to the veterans hospital to die. Having read about the work of faith healer Kathryn Kuhlman, he attended a service she conducted after which he simply threw away his crutches and walked unaided. Readers who have attended workshops conducted by Dr. Reams know he would begin each day with a devotional talk.

Dr. Reams worked as a consultant to agriculture where he developed his ideas pertaining to raising high-"energy," high-brix crops via balancing soil mineral content, improving the carbon content, and providing foliar, side dress and other supplements to the soil. He conducted seminars and tested soils under his own company, International Ag Laboratories and, later in life, he teamed up with Dr. Dan Skow to hold seminars for farmers and consultants.

In addition to his agricultural endeavors, Dr. Reams operated several clinics, but his skills in utilizing diet and food supplements for the benefit of people seeking health soon made him the target of the medical establishment. According to interviews in *Acres U.S.A.,* Dr. Reams earned a doctor of science degree plus held a medical degree from England.

He is known for many contributions, among which are: a unique urine and saliva testing system to evaluate human health; use of the "Reams" approach to testing both human and soil samples (Morgan or universal extraction solution to measure available nutrients); the concept of plant/animal growth resulting from "energy" and not simply food/fertilizer; the importance of calcium in the diet of animals and soil and how this calcium reacts with other elements and nutrients to produce energy; the importance of phosphorous as the "carrier" of elements into the cell required to build a healthy cell (vs. nitrogen as a carrier which builds a watery cell); the use of the refractometer to measure plant "health"; the emphasis on calcium, sulfate, phosphate and carbon in contrast to nitrogen and potash in growing crops; the understanding of "growth" and "fruiting" and how plants can be manipulated.

Chapter 2

Energy

Simplification and reductionism are just not nature's way. Energy is a concept seldom talked about by growers, yet its ramifications are crucial to a current understanding of today's agriculture. Dr. Reams stressed one crucial point in all his lectures: plants grow from energy, not from fertilizers. He stated that different fertilizers, even of the same grade, would have different energy levels depending upon the ingredients of the fertilizer and the filler content. This is why two fertilizers of the same grade (Brand X 10-10-10 vs. Brand Y 10-10-10) will perform differently in the same field. It is also the reason why one manufacturer's product out-performs another despite the claim they both have "equal amounts of plant food."

Energy Content

Dr. Reams taught that the energy content of any given fertilizer or chemical could be calculated by using a mathematical formula. In using his calculations, one can determine that the energy of a single atom of calcium may range from a low of 540 millhouse units to a high of 20,959 millhouse units. Correspondingly, a single molecule of calcium carbonate (high-calcium lime) ranges from a low of 30,544 to a high of 82,895 millhouse units. This range is obviously quite extensive. With this understanding, it is easy to see how a product from one supplier responds in the soil very differently from supposedly the same product obtained from another supplier.

This fact has been confirmed by farmers on countless occasions. Energy differences account for one bale of hay providing more feeding contentment than three bales from another field. These same energy differences account for one seed variety out-performing another on the same plot.

Energy Signals

This concept has been confirmed by Dr. Phil Callahan, retired USDA scientist from Gainesville, Florida. Dr. Callahan performed pioneering work in the detection of subtle energy signals given off by plants, insects, and even rocks. Through his experimentation, he discovered that insects use their antennae to detect and identify energy signals from plant emissions. These signals identify the plant source as well as the plant condition; e.g., healthy or non-healthy alfalfa.

The understanding of the concept of energy is crucial. Every atom and/or molecule contains energy. The energy in the atom reflects the energy in the plant, animal, or product. Energy is required in the movement of anything from one place or state of matter or form to another. Energy is released in the utilization of fuel whether it be petroleum, fertilizer or food. Different fuels may contain the same label regarding the contents; e.g., gasoline, alfalfa, or 11-52-0; however, the astute observer knows that when products originate from different sources, they often perform differently. Calcium carbonate, or high-calcium lime, will vary in energy depending upon the mine it came from. With our current understanding and technology, including the use of an electronic soil tester (see Chapter 6), we are better prepared to make use of this information. The capacity to measure these energies has not been available until recently. Farmers have known about them based upon their own observations and experiences, so the concept of energy *per se* has not been at issue. Rather, explanations of energy and predictions regarding their cause and effect have been questioned.

From the early days, we have been taught that "likes repel and opposites attract." Simple experiments with magnets and electricity show this principle. This theory does not always apply, however. If likes repel, how can a vein of gold, silver, or copper develop from molten magma? In many cases, likes attract.

Energy Creation

According to Dr. Reams, energy is created in a soil by the interaction between nutrients — sulfate, phosphate, etc. — and calcium. Calcium is the driving force required. As the other nutrients interact with the calcium, energy is released which powers the growth of the plant. This energy creation also is driven by the monthly moon cycles. As the days of the month cycle, soil energy fields are replenished due both to the bathing of the earth by cosmic energy fields so eloquently described by Carl Sagan, as well as to the magnetic and gravitational attraction of the moon.

Energy Field

Dr. Reams taught that when it comes to agronomy, likes do attract. Putting small amounts of high-energy fertilizers on a field, especially in the fall of the year, can result in dramatic increases of the same materials appearing on soil tests in the spring. This can be attributed to a synergistic/attractive force that may either release like materials in the surface soil, draw up like material from greater depths, or increase (strengthen) the energy signal of the base material in the soil.

Fertilizers in a liquid medium are excellent for creating such an attraction or strengthening because the fertilizer molecules can cover so much more soil when dissolved and dispersed in water. Using 20 gallon nozzles, much more soil will come into contact with the fertilizer applied as liquid than its dry counterpart. Even admitting the dilution factor, a small amount of fertilizer *energy* comes in contact with soil and, therefore, increases the total force which attracts similar energies. Thus, broadcast spraying only 10 pounds of 12-50-0 dry soluble fertilizer can sometimes provide adequate phosphate energy in a balanced soil to attract or release sufficient amounts of phosphate energy for successful crop use.

Magnetic Energy

Another very important form of energy involved in plant growth is magnetic energy. Two distinct phenomena seem to be functioning in a healthy soil. The first was described by Dr. Reams and concerns movement of soil nutrients along the earth's surface following the south to north magnetic flow. This magnetic stream carries ionized minerals along its flow. These minerals are then deposited onto, or are attracted to, plants and/or soil along the path. The magnetic effect also functions above the soil, and plants will be poor or better *antennae* according to their nutritive state. Dr. Callahan expands this concept by explaining monopole or single pole magnetic attraction. One monopole enters the plant directly, and the opposite type enters rocks or soils and then travels into a plant. The meeting of the two opposite monopoles in the plant releases energy.

A third phenomenon is the paramagnetism of the soil itself. A healthy, fertile soil is slightly attracted to a magnet and, therefore, is considered paramagnetic. This is a natural state of matter and is not the same as ferro-magnetism which explains the attraction of metal to a magnet. Recent research by one of the authors, Dr. Wheeler, has shown that the majority of traditional soil-building materials, such as compost, granite dusts, and colloidal phosphates are only slightly paramagnetic. Most dry salt fertilizers are very slightly paramagnetic. Certain dry-soluble fertilizers (see Chapter 10) appear to be diamagnetic (repelled by a magnet) which may be one factor in their efficiency. The mono-ammonium phosphate molecule may be magnetically significant in fertilization which may account for the relatively good performance of such standard fertilizers as 10-34-0 and 11-52-0. Readers should consult the bibliography for more information on the above concepts.

Chapter 3
The Soil Environment

To farm successfully for the long term, the farmer needs to work with living soil. Soil was created as a living mass made up of millions of living organisms each complementing the other. A September, 1984 *National Geographic* article revealed how soil microbes have certain roles to fulfill within this complementary arrangement. Microscopic in size, microbes number in the billions and populate all healthy, living soils.

Microbes require a delicate balance of soil tilth in order to live and survive. An ideal condition would provide for 5 percent humus, 45 percent mineral, 25 percent air, and 25 percent water. A soil with this profile would be fairly easy to plow or till, would be spongy to walk on, would hold large amounts of rainfall with minimal erosion, and would provide air (oxygen) needed by aerobic microorganisms.

The type of microbe that flourishes in a given soil is also highly dependent on the plant life growing there. The root zone microbial types and populations are plant dependent, as reported in Russian research of the 1930s and 1940s. Such a strong symbiotic relationship exists between plants and soil that it could be argued that the plant exists to build soil rather than the soil being used to grow plants.

Microorganisms

The term *microorganisms* covers a broad array of live organisms including bacteria, actinomycetes, fungi, algae, protozoa, and nematodes. The decomposers play the crucial role of breaking down organic waste into usable nutrients which are then chelated or locked into the resulting humus. Some organisms partially break down organic wastes, e.g., corn stubble and manure, while others consume these breakdowns taking the decomposition process one step further. Completely broken down nutrients may be stored on clay and humus factions of soil which can then be used by plants. Other microbe by-products such as vitamins, amino acids, enzymes, hormones and antibiotics can aid in plant growth. Within this environment, plants sprout and grow to their optimum limits according to the available energy and nutrients they are able to absorb. They seldom reach their genetic potential due to the limits of agricultural systems, weather, and our lack of knowledge of how to tap into their full potential.

Organic Matter and Humus

Manures and other animal or plant remains serve to replace mineral and protoplasmic matter directly to the soil. In both instances, they are potential plant food which must first be composted (consumed by microorganisms) before they are available to the plant. Once composting has occurred, the material is considered almost 100 percent plant food.

Soil microbes break down plant and other remains in a highly cooperative fashion. No one bacteria or fungi accomplishes the task alone, but, in concert, all cooperate to accomplish the end result. Plant food originating from such a source usually is rather stable in the soil, doesn't leach out and, in fact, is the primary reason why soil is able to hold water. This material is called humus.

Humus differs from organic matter in that it is the final result after bacteria have digested or composted the organic matter. The identity of the plant materials is lost in the humus form. The rapid proliferation of bacteria in a soil due to the addition of carbohydrates or mineral humates has sometimes changed the humus content very rapidly and caused learned men to claim that it is impossible to do in the time stated. Humus and dead bodies of microbes are very difficult to distinguish from each other and really constitute the total humus content. Parts of the total humus are highly available to microbes for their own food or direct uptake by plants while other parts resist breakdown under all but the most harsh substances such as anhydrous ammonia or fumigants.

With the recent emphasis on "cash crop" farming (farming which grows only crops to sell for cash rather than to use as feed for livestock), farmers have seriously neglected humus levels. They no longer have animal manures to spread back on the fields, and they often fail to add organic matter back in the form of cover crops. Consequently, soil organic matter and humus levels have decreased dramatically on conventional farms in the past 50 years.

"Active" Carbon

Carbon comes in many forms. The diamond in a wedding ring and the coal mined in Kentucky are both forms of carbon. For agricultural purposes, the active form would be humus. "Active" means it is biologically beneficial to soil microorganisms and assists in the building of soil life. Active carbon in the form of humus provides three major contributions.

The first contribution is that active carbon can hold four times its weight in water. The lack of active carbon (good humus) is the primary reason why farmland is eroding. Soils with a humus level of 1 percent will hold approximately 10,000 gallons of water per acre. A one-inch rain supplies 28,000 gallons of water per acre. If a field with 1 percent humus receives a one-inch rain, it can easily become super saturated and erode. Soils with 6 percent humus levels will hold an excess of two inches of water. Thus, a two-inch rain on a 6 percent humus field will readily be absorbed and held with a minimum of erosion.

When farmers remove every cutting of alfalfa or chop corn for the silo, they are returning little organic matter to the soil.

The alfalfa farmer is returning nothing while the corn farmer is returning only the root mass developed during the year. This is poor organic matter practice, and it is why recent emphasis has been given to growing cover crops which will at least provide a green manure to return to the soil. A good suggestion would be to cut and leave the last crop of an alfalfa field each fall as an additional humus builder or apply manures.

Active carbon is crucial not only because of its water-holding capacity, but also because it helps "fix" nutrients in the soil and buffers salts. When bacteria and other soil life are destroyed through salt fertilizers, toxic chemicals or compaction, their reduced numbers greatly affect organic matter decomposition. Improper decay often results in the formation of formaldehyde and other aldehydes which are toxic to the soil. This explains why corn stalks can be plowed up completely preserved after being buried for two years. Thus, a vicious downward cycle has been started.

Poor microbial life results in improper decay that further reduces the organic life which then yields more toxic by-products. This leads to the need for chemical fertilizers to provide plant fertility needs which are not being met by an active soil environment. A recent study by Reganold and Âlliot shows 6 inches more topsoil remained over the check field during 40 years of using biological promoting methods compared to standard (acid/salt) fertilizers.

The third contribution provided by active carbon is that of regulating the magnetic flow across the field and the general energy pattern throughout the field. Active carbon holds or stabilizes field energy. When the magnetic energy system is working properly, soluble nutrients flow at the correct speed to interact with plant root hairs. Too fast or too slow a speed can result in apparent toxicities or shortages of given nutrients. This "speed" concept can be related to pH.

Magnetic Flow and pH

Since energy is the key to crop production, it is important to provide energy to the field as well as to create conditions in the field whereby energies from the cosmos, fertilizers, rain, sunlight, etc. can be received, controlled, and transferred. This is done through the use of cultural practices and soil additives which can create conditions for the energy system to function efficiently.

Although pH is usually thought of as a measurement of acid or alkaline properties, it can also be thought of as a measurement of energy flow. This "energy" flow definition is helpful in understanding pH for farming applications. When the pH is too low (acid) relative to the type of crop, the energy flows too rapidly. Nutrients literally pass by plant roots too fast to be properly attracted to and absorbed by the root. Some heavier metals, like manganese, may be attracted to the plant too quickly, and manganese toxicity may result. Alkaline soils, may have a slower than optimal energy flow. Nutrients travel too slowly to interact with hair roots, and the crop may exhibit nutrient deficiencies and may not produce to its field potential.

Chapter 4
Weeds

Weeds play a vital role in the soil. One definition of a weed is a plant growing out of place. A corn plant is considered a "weed" in a soybean field. If only we could find a market for foxtail, ragweed, or velvetleaf . . . . Could it be that weeds play an important role in soil ecology aside from covering the soil?

Why Weeds Grow

A closer look at weeds reveals that each weed species has a preferred soil type or condition in which it likes to grow. Pigweed, for example, grows in compacted soils with relatively high fertility. Joseph Cocannouer in *Weeds, Guardians of the Soil* explains that weeds actually perform valuable services. Weeds such as pigweed and lambsquarters have strong taproots which push down and open up the soil. Water and minerals are then carried up alongside the weed root to the topsoil where the small feeder roots can then use them.

Feeder hairs are very delicate. They require soft, aerated soil and will not survive in hard, dry soil. Often weeds will benefit a corn or other crop by making water and minerals available which would not otherwise be accessible.

Upon careful examination, Cocannouer found weeds to follow certain growth patterns. For instance, weeds will often contain high amounts of the particular mineral which the soil is lacking. This was also confirmed by Dr. Wheeler early in his career when he submitted weeds for tissue analysis and then compared them to soil test results from the same field. Understanding or knowing why a certain weed grows can give the farmer a clue as to how to go about relieving the pressure from that kind of weed.

Among the benefits of weeds are:

1. They are deep rooted and can store minerals and nutrients.

2. They bring minerals, especially those depleted, up to the topsoil and make them available to crops.

3. They fiberize and aerate the soil.

4. They help break up hardpan.

5. They are good indicators of soil and mineral conditions.

In his book, *Weeds!!! Why?*, Jay McCaman makes the statement, "Weed populations can be better indicators of mineral availability than soil tests!" In his expanded book, *Weeds and Why They Grow*, he lists more than 800 weeds and gives the soil characteristics associated with their growth. According to McCaman, foxtail, nutgrass, and water-grass grow where there is an excess of carbon dioxide (often caused by compaction and/or poor levels of aerobic bacteria). The same conditions will promote corn smut and corn rootworm. Dolomite lime will contribute to these conditions. "Where magnesium is elevated," he says, "look for foxtail and fall panicum. In soils low in calcium (often, but not always, low soil pH), look for field bindweed, dandelion, nightshade, redroot pigweed, and lambsquarters."

Poor Soil Decay

McCaman says, "Poor bacterial action will cause improper decay." In soils high in methane gas, he states, look for velvetleaf and jimsonweed. Soils high in ethane also have poor decay. Lack of soil fungi will mean rhizomes will not be digested and weed growth will not be biologically controlled. Redroot pigweed grows in compacted soils and sandburs grow in soils low in humic acids. Mustard grows where salts are present or being released. Broadleaf weed pressures can often be controlled by balancing phosphate and potash ratios. As the available nutrient ratios drift from 2:1 (P:K), broadleaf weed pressures will increase. When the ratio becomes close to 1:8, weed pressure can become so intense that herbicides may no longer be effective.

Grasses can be brought under control by raising biologically-active calcium levels. High-calcium lime and liquid calcium are excellent ways of raising calcium levels. Liquid calcium and molasses mixtures have been used to suppress weed germination when applied immediately after planting. Many farmers have seen weed control using this approach which equaled the results of toxic chemical programs.

Many weeds have become resistant to herbicide applications developing into so-called "super weeds." Although herbicides may cause soil damage, much of the current day weed problems have been caused by the improper use of poor quality fertilizers that have damaged and imbalanced soils. By following the conventional recommendations to use acid, salt and chloride fertilizers, the farmer has seen his weed and grass problems increase year after year.

Eliminating the Need for Toxic Chemicals

For farmers wanting to try a new fertility approach and reduce or eliminate their use of toxic chemicals, four steps are required. First, recognize that toxic chemicals are, in fact, toxic. Although they may be given fancy marketing names that suggest power or success, their generic classification always contains the suffix -cide as in pesticide which means to kill. They contain killing vibrations or "energy" which work not only on weeds, soil and soil life, but on ground water, the farmer, and his livestock as well. Their whole purpose is to kill some living thing and, consequently, they are dangerous to use. Not only do they require caution in mixing and applying, but once applied, they can combine with other chemicals or break down into new compounds in the soil and result in new molecular structures with potential toxicity.

These newly combined chemicals have not been identified or researched, and we are only now becoming aware of their accumulation and potential danger.

Second, experiment with reducing the amount of chemicals used. Test plots used to determine safe application rates always show a range of use effectiveness. This range may indicate poor weed kill at 1 quart per acre to excellent weed kill at 6 quarts per acre. By recognizing that this range exists, the farmer can use the lower limits, be satisfied with a "moderate kill," and know he is working to move off the product in the next year or so. There are many additives that can be used with herbicides to increase their effectiveness at lower rates. These include such products as soybean oil, nitrogen, liquid calciums, garlic, wetting agents, etc. He can also start banding over the row and use cultivation between the rows. As the soil recovers, the weed pressure will also recover.

Third, build up or balance soil fertility to eliminate the primary cause of the weed problem. Weeds grow in response to soil imbalances. Once these imbalances are corrected, the conditions supporting this weed population no longer exist. By changing the soil conditions, the farmer will change his weed patterns. This is best done via the use of soil tests (see Chapter 6). This approach does not necessarily mean the need to purchase extensive amounts of fertilizer. Often small amounts of higher energy fertilizers (see Chapter 10 on the use of bio-activators) will produce excellent results.

An Amish farmer in Pennsylvania was confronted with a pasture overrun by spiny Jimson weed. Foliar spraying two gallons each of liquid calcium and black strap molasses, eight ounces of soil conditioner and 12 pounds of a dry soluble fertilizer almost completely eliminated the problem.

Finally, realize that Mother Earth will attempt to recover herself once the farmer reduces, changes, or eliminates his problematic fertility and chemical applications. Given the chance, nature will degrade toxic materials, balance nutrient levels and pH by microbial activity and return productivity.

Weeds, like insects, are excellent indicators of imbalances. Once you recognize them as being a help, you can take steps to modify your fertility approach, correct soil imbalances and eliminate your weed pressures. Weeds are messengers. Although our initial impulse may be to kill the messenger bearing bad news, a more constructive approach may be to listen to the messenger and respond to the cause itself.

Chapter 5

Insects

Insect Damage

Insects and insect damage have been called the "farmer's curse." It is true that each year millions of tons of produce, grains, and fruits are destroyed or damaged by insects. Insects account for a 13-16 percent loss from $244 billion in crops annually in the United States. Insect numbers count in the billions and their collective weight by far surpasses the collective weight of mammals. Of more than a million zoological life forms identified and categorized by scientists, more than 800,000 consist of insects. It is believed that as many as 10 million insects remain as yet to be identified. Aside from our annoyance with these pesty critters and their attacks upon crops, pets, and livestock, what is their purpose?

Insects actually benefit man. Estimates of the value of insect pollination from honey bees and wild bees alone amount to approximately $30 billion annually in the United States. Insects pollinate fruits, berries, grapes, and field crops including peas, onions, carrots, clover, alfalfa, and flowers. In addition, insects provide millions of dollars annually in the form of such items as honey, shellac, and silk. Many insects are actually beneficial to man because they devour insects harmful to our crops. Ladybugs, for example, will eat aphids. These predators play a useful role in maintaining balance within the insect kingdom.

Less than 1 percent of the insect species are considered harmful. About 1,000 species are considered serious crop pests, another 30,000 species are described as minor crop pests. Their control cost is only slightly less than the value of the crops they would have destroyed if left alone. In 1995, worldwide expenditures for pesticides hit $37.7 billion; U.S. expenditures came in at $11.3 billion.

Conventional Control

Insecticides are the modern mode of insect control. Insecticides come in either dry or liquid form and are either dusted or sprayed. They are used to prevent insect damage as well as to kill the insects after they have arrived. Insecticides come in several types. Some are stomach poisons which react within the insect after being consumed. Others kill on contact. Others, called systemics, are absorbed by the plant or animal and affect the insect after it bites the treated host.

Now that public awareness has increased and public opinion has caused the EPA to review pesticides, it is expected that many will not be allowed to remain on the market. This scenario has prompted Steve Brown, Auburn University Extension Service, to list several alternatives for farmers to consider. These can be considered as part of an IPM or Integrated Pest Management program.

a. Select insect-resistant varieties.

b. Calculate closely such variables as planting dates and row spacing.

c. Take advantage of crop rotation benefits.

d. Utilize pheromones (insect sex attractants) to capture or disrupt insects or introduce predator insects.

e. Utilize the biological pesticides which are available.

f. Consider trap crops in certain instances.

g.   Utilize plastic mulch.

h.   Consider soil solarization, using clear plastic.

i.   Utilize machinery which sucks insects off plants.

Although these suggestions represent creative solutions to a growing reality, they miss the mark in that they don't address the cause for the insect infestation in the first place. Once the variables influencing insect attack are understood, steps can be taken to remedy these causes. Addressing the cause will produce more lasting results.

Infrared Signals

Dr. Philip Callahan, renowned authority on the corn earworm and author of *The Soul of the Ghost Moth* and numerous other books, has studied insects extensively in his role as USDA researcher. His research indicates that insects communicate via infrared signals which are received and sent by the insect antennae which occur over much of their bodies. Each insect is apparently sensitive to certain plant signals and ignores others. Most damaging insects are selective in what they attack. Thus, the alfalfa weevil would not infest elm trees.

Infrared signals are emitted naturally by all living plant or animal bodies as well as from the gaseous emissions of all plant and animal life. Signal strength and configuration are affected by a variety of factors including nutrient balance and stress factors. Insects detect these signals with their antennae.

Upon close examination, it is evident that each species of insect has an antenna shape unique to its species. According to Dr. Callahan, the shape of the antenna determines the signal range received by the insect. Thus, the shape of weevil antenna allows it to be attracted to alfalfa frequencies.

When plants are grown in a soil with balanced nutrients and the plant itself utilizes those nutrients in a balanced manner, its own system will maximize its genetic potential in terms of yield and health (or resistance to stress). However, when the soil is out of balance, when normal growth stresses, e.g., drought, excess water, heat or cold, wind or hail occur, the plant may require other nutrients to counteract the stresses at hand. The extent those nutrients are missing is the extent the plant will suffer and, eventually, deviate from its genetic potential. The infrared signals given off by the plant will modify depending upon the health of the plant. As the plant moves further from ideal health, the signals become more pronounced in a way that attracts insects. This can be shown by taking refractometer readings and observing that the brix reading measured as percent sucrose on attacked plants is lower than plants not being attacked. The brix reading is a good indication of the efficiency of the plants' output of carbohydrates which is the result of photosynthesis.

Soil Balance-Imbalance

A properly balanced soil will have sufficient quantities of organically active carbon — humus — which helps hold nitrogen in the ammoniacal form. In soils lacking this active carbon content, the soil will give up this ammoniacal nitrogen to bacterial conversion into nitrates or directly to the atmosphere in gaseous form. During the process of ammoniacal nitrogen leaving the soil, it passes by the plant and can act as an amplifier of the infrared signal coming from the plant. Whereas the plant may have been initially broadcasting the signal, "I'm not balanced nutritionally," the signal now reads, "Come and feed on me!"

Dr. Reams taught that most insects do not attack healthy plants. His whole approach to plant fertility and insect control capitalized on supplying the soil balanced forms of plant food which, in turn, maximized plant health. Insects look for signals coming from unhealthy plants and seldom attack healthy ones. Insects willingly eat weeds and will return to that practice in fields with healthy crops and soils and unhealthy (low brix) weeds. The attacking of weeds by insects is one of the signs to look for in observing your progress toward sustainable agriculture.

Failing Plant Health

The research conducted by Dr. Callahan and Dr. Reams has immense implications. If insects attack unhealthy plants and ignore healthy plants, they are telling a sad story about the fertility approaches as currently practiced. By attacking unhealthy plants, insects are actually benefiting humanity by pointing out which plants are unhealthy, low in mineral content, and not fit for human or animal use. The astute farmer views insects, as he views weeds, as messengers of soil or crop conditions, not the cause of them.

Natural Control

Many farmers are beginning to work with the IPM (Integrated Pest Management) approach to insect control. President Clinton once announced his intention to have a large percentage of USA crops grown under IPM by the year 2000 in an effort to reduce the amount of toxic chemicals used. This concept consists of setting out insect traps baited with the sex scent (pheromones) of insects and then observing insect populations. If the insects are present, but in a number below that which would cause significant crop damage, no spraying should occur. If the population indicates significant crop damage will occur, steps are taken to control their numbers, hopefully with non-toxic materials. Other aspects of IPM include the release of mating disruption pheromones or predator insects to devour the harmful ones present on the crop.

Increasingly, farmers are turning to non-synthetic pesticide options such as botanical, microbial or predator approaches. These consist of using plant extracts such as nicotine from tobacco leaves, pyrethrum from flowers, rotenone from roots as natural insecticides; using plant extracts such as garlic juice and capsicum from peppers as repellents; microbial vectors that destroy harmful microbes or larger organisms; and predatory insects to control insect pressures. Ladybugs and lacewings are traditionally welcomed in the field as a predator of moths and other destructive insects. Additionally, their presence usually indicates a relatively low level of toxic contamination in the field, since they are also killed off by toxic sprays. Ladybugs usually are considered an indication that the field environment can sustain beneficial insect life.

It is important to consider using a foliar nutrient or feed with any type of insecticide whether synthetic or natural. Any plant under attack by insects is mobilizing its defenses. This requires nutrient and energy utilization. Wouldn't it be wise to give some "chicken soup" to your crop along with anti-insect treatment to aid in its recovery?

An interesting natural product for insect control is diatomaceous earth. D.E., as it is commonly called, consists of the shells of tiny fresh or sea water diatoms which have been deposited on old lake beds over millions of years. They are mined and milled into powders for feed or for use as a filtering agent in swimming pools. The swimming pool product cannot be used in feed as it will damage the animal consuming it. Since it will absorb many times its weight in water, D.E. is considered to be an anti-caking ingredient for feed. It is often fed by alternative ag farmers, not because of its anti-caking properties, but because of claims it will control parasites in animals. Although it feels like talcum powder to the touch, you would see extremely sharp edges under a microscope. Supposedly, when the substance comes into contact with an insect it will scratch the insect's cuticle. Death often follows from dehydration. How it works internally is not fully understood. Some think it *de-energizes* the parasite in the stomach.

Although only a few brands of D.E. on the market have gone through the EPA registration requirement to be considered a pesticide, other brands could work the same. Recent university research has shown that the vegetable oils used with pesticides may also give excellent insect control when used alone. However, the EPA has yet to "catch up" with this information and give its full "blessing."

Could it be that insects and weeds are symptoms of a problem rather than problems themselves? Could it be that fertility approaches exist which can correct these basic problems exemplified by insect and weed pressures? Are these pressures related to fertility practices? If this is the case, how does the farmer determine the correct fertility program to use?

Chapter 6
Soil Testing

Taking A Soil Sample

The taking of a soil sample is almost as important as the test result you get back. You'll need a very clean shovel (rust shows up as iron) or better yet a good stainless steel soil probe. Take frequent probes throughout the field. Choose the depth or horizon according to what you are trying to learn. Usually you are interested in the root zone where active absorption or use of nutrients is occurring. This zone should be aerobic and is usually sampled at 0 to 6 inches for annuals and 0 to 8 or 10 inches for perennials. For no-till operations, it is wise to sample several 2- to 3-inch horizons starting at the soil surface and continuing to at least 6 inches for annuals or 8 to 10 inches for perennials. Many labs recommend taking one sample for every 15 to 20 acres with a minimum of one probe per acre.

Place the soil in a clean plastic pail. (The story goes that one grower took ten soil samples using a plastic pail which had been used to measure feed ingredients. All except the last sample had very high levels of sodium.) After all the cores are taken mix thoroughly, and send a pint measure to each lab you want to perform a test.

Remember, you will be sending a soil testing lab about eight ounces of soil from which you will make fertility programming decisions . . . just eight ounces compared to 2,000,000 pounds of soil per acre. Taking a sample from too large an area leaves you with poor information from which to consider. How many of your fields have uniform soil types throughout? How many have a variety of soil types? How will you be treating the field being tested? Will you fertilize it the same? Will you be treating each soil type differently?

Growers usually can visually see areas within the field which grow more poorly than the remaining field. If these are large enough to treat separately, sample them separately and fertilize accordingly. If they will not be fertilized separately, skip them all together. Better to have a small plot which does poorly than to skew the results for the remainder of the field. These questions give an indication as to the complexity of testing soils. The authors have seen a gridwork of a field in which soil testing was performed by the University of Minnesota on 5-meter squares. A separate gridwork was shown for each of the nutrients, e.g. N, P and K. Extensive variations existed throughout the field. How could a farmer afford to test as extensively as this? How could he be assured that he sampled from the right places?

Soil testing is done not only to provide data from which to make fertility programming, but also to be able to monitor progress over the years. If you sample from one location on one occasion and from a second location on the second test (and, unless you placed a marker how would you not?), it is difficult to know with exact precision if progress is being made. If the two locations sampled had very different fertility make-ups to begin with, how could subsequent test data show progress?

Questions like these are not posed to discredit the soil testing industry, but rather to give an indication as to how complex the soil testing concept is. Ideally you would sample from year to year from the exact same location. The new GPS technology, which will allow the farmer to locate, often within a few inches, the same field site, will greatly help this field monitoring process. Farmers who walk their fields, experiment with various approaches and keep good records, and who observe which fertility programs work will generally be in the "ball park." But to get to "first base," soil testing is required. Three different types of soil testing will be discussed.

CEC Test

The late William Albrecht, Ph.D., Professor of Soils and Chairman of the Department of Soils at the University of Missouri College of Agriculture, was the father of the Cation Exchange Capacity (CEC) test used by most commercial and university soil labs. This test measures the "holding capacity" of soil and determines how much nutrient is theoretically being held by the clay and humus colloids.

According to the CEC theory, clay and humus are negatively charged and will "hold" positively charged minerals or soil nutrients. The greater the percentage of clay or humus, the greater the holding capacity of the soil. Sandy soils have relatively little holding capacity and, therefore, need to be fertilized and watered frequently to replace the nutrients and water that leach out. Soils can hold more nutrients with the accumulation of humus and/or clay. This increases their CEC reading and changes their classification to sandy loams, clay loams, etc. Muck or peat soils, composed largely of organic matter, have the greatest holding capacity. Sandy soils will have a CEC rating of 0 to 4, loamy soils of 5 to 8, loamy clay soils of 9 to 15, while heavy clay, muck or peat soils of from 16 to 30 or higher.

The CEC test uses chemical solutions to extract nutrients, e.g., calcium, potash, and magnesium, from the soil being tested. An analysis of the results will provide a percentage of the holding capacity actually filled with the minerals being tested. Many CEC laboratories recommend the following percentages as being appropriate:

$$\begin{aligned}
\text{calcium} &= 60\text{-}70\% \\
\text{magnesium} &= 12\text{-}15\% \\
\text{potassium} &= 2\text{-}5\% \\
\text{sodium} &= 1\text{-}2\%
\end{aligned}$$

By determining the current percentages in the soil, the farmer can decide which nutrients need to be added to balance his soil mineral content and to produce the best yield as well as soil conditions.

Dr. Reams felt, however, that whereas this testing approach may have given adequate recommendation results in years past, it doesn't seem to be as effective today. He felt that this test is better used for long-range fertility planning than for determining nutrients needed for the immediate growing season. The test gives an indication of how much of a nutrient is in the "savings account," but not how much is in the "checking account." By "savings account," we're referring to nutrients held in reserve in the soil, whereas, nutrients currently available for plant use are categorized as being in the "checking account." The variation between the two "accounts" is an important factor in determining the biological availability of the nutrients which is also dependent on soil life.

LaMotte Test

The LaMotte Chemical Company of Maryland has developed a water-soluble testing approach which some prefer to use in addition to the CEC test. The LaMotte procedure uses solutions for nutrient extraction which, supposedly, are more similar to those produced by the plant roots. The plant extrudes mildly acidic solutions to dissolve minerals, and the LaMotte test extraction solution supposedly resembles these natural solutions.

Dr. Reams was a strong advocate of this water-soluble testing approach. He encouraged farmers to purchase testing kits and perform tests throughout the year. By measuring which soil nutrients were actually available to the plant twice a month, the farmer could decide what nutrients were necessary to add to the soil as the spring and summer months progressed. Dr. Reams recommended the following levels of water soluble nutrients in the soil solution on a per acre basis:

ammoniacal nitrogen	40 pounds
nitrate nitrogen	40 pounds
phosphate	200-400 pounds
potash	100-200 pounds
calcium	2000-8000 pounds
magnesium	14% of calcium level
sulfur	200 pounds
manganese	30 pounds
iron	40 pounds
zinc	15 pounds
copper	15 pounds
boron	3-5 pounds
chlorine	3 pounds

The LaMotte testing approach can be performed by the farmer if he is willing to learn the test procedure and to take the time required to perform the test. Dr. Reams recommended that monthly or bimonthly tests be taken in order to keep the soil solution "current." This frequency of testing makes economic and management sense for rapid-growing, high-value crops such as vegetables, fruits and flowers. The need for "in lab" tests can be reduced by "in field" monitoring as well. Many soil consultants recommend the CEC and the LaMotte tests be used in combination.

They suggest the CEC test be used as a "mining assay," meaning that the test measures those minerals to be found in the soil if the soil were "mined." The mining assay tells what is actually in the soil and not necessarily what is available to the plant. They then use the LaMotte test to determine which nutrients are available to the plant and in what amounts. For example, if the CEC test shows low calcium levels, they would recommend applying high-calcium lime to raise the calcium levels in the soil. If the CEC showed medium to high levels of calcium and the LaMotte test showed low available levels, they would recommend a liquid calcium or possibly a liquid sulfate compound to make the calcium available to the plant. They might also recommend a bacteria-enhancing or live bacteria product to enhance the soil bacterial action in breaking down the lime which then would make calcium available.

Electronic Scanner

In the 1920s, experimentation was being undertaken on a device subsequently called an electronic scanner which measures subtle energy patterns. Dr. T. Galen Hieronymus, an American electrical engineer, developed what he called an *eloptic energy analyzer*. His work resulted in English, U.S., and Canadian patents issued around 1950. Others have developed similar devices. The device consists of a box containing electronic capacitors, tuning dials, and switches.

Dial rates for all elements — calcium, phosphorus, nitrogen, humus, etc. — are available. The operator simply dials each of these rates into the machine and determines the energy intensity for each element being tested. Once energy readings are obtained, the operator draws up a composite reading of the field in question and can make additional determinations. Having checked a sample of the farmer's soil, the tester can then check the energy intensities from seed samples being considered by the farmer for planting. The operator can choose the best seed to be planted in the soil. Additionally, the operator can select the best fertility program for this seed-soil combination. Since each fertilizer brand differs in energy from its competitors because of the source and quality of the raw materials going into the fertilizer, these energy differences can be measured via the electronic scanner. By testing a variety of fertilizer products, the operator can build a fertility program which best matches each field/seed combination. Using this approach, the tester can build up the vitality (potential growth energies) in the soil or crop while, at the same time, building fertility programs which reduce the vitality of the weed or insect problems the farmer is encountering. The programs may or may not contain conventional toxic chemicals.

The results obtained from competent operators have been impressive. Its use has helped farmers cut input costs, increased yields, and produced a higher quality crop. Higher quality is especially important to farmers growing feed for their livestock.

Electronic Scanner Results of Some Soil Applied Products on Selected Weeds, Insects and Crops Plants or Insects

Product Tested	Quack-Grass	Velvetleaf	Redroot Pigweed	Corn Earworm	Alfalfa Weevil	Hybrid Corn	Alfalfa	Wheat
Vitality (G.V.)	130	470	160	320	300	450	480	310
Soil Conditioner	10	230	10	20	60	600	660	420
Liquid Humic Acid	0	100	10	20	30	690	1,100	720
21-0-24S	20	40	10	30	10	630	610	470
9-18-9	10	230	20	30	30	1,000	590	500
Anhydrous	270	820	240	510	1,200	190	20	80
28% N	20	170	30	60	30	840	410	370
Colloidal Phosphate	0	10	10	0	0	4,600	4,700	820
10-34-0	30	30	20	0	20	670	610	560
11-52-0 (MAP)	170	60	20	10	30	630	500	410
0-46-0	110	360	140	80	150	250	190	170
0-0-60	70	740	250	640	740	120	200	100
0-0-50	50	300	180	60	60	600	370	500
Dolomite Lime	230	100	160	300	50	1,000	770	610
Hi-Cal Lime	0	0	0	40	30	3,100	5,400	2,200
Gypsum	0	10	20	60	10	490	1,100	530
Liquid Calcium	0	70	20	30	10	1,200	3,300	710

Indirectly, it is also important to the general public who consumes food directly in the form of corn flakes, breads, and produce.

An example of what can be determined using an electronic scanner is found on the previous page. Keep in mind that the readings indicate energy intensities, not pounds per acre.

Thanks to Jay McCaman for his permission to reprint this data from his book, *Weeds!!! Why?* These figures were obtained using an electronic scanner to test products available from his locale in Michigan.

Understanding Energy

To interpret this information, the term *vitality* needs to be explained. Vitality is used to describe the intensity of an energy reading obtained from analyzing the soil, product, weed, insect, or crop in question. Every soil, plant, or product emits an energy pattern peculiar to itself. This pattern can be measured, and its intensity compared with intensities from other products or plants.

Using the scanner, the operator compares the energy intensities from plants and/or soils to those of products. If the resulting plant or soil electronic energy pattern intensity is greater, the product can be said to "raise" the vitality of the plant (or soil). If the result is less, the product has "lowered" the vitality of the plant (or soil). The farmer is obviously interested in raising the vitality of the crops he is growing and lowering the vitality of the weeds and insects he is trying to combat. Note that quackgrass on the previous chart has a vitality of 130. Notice that corn earworm has a vitality of 320 while hybrid corn has a vitality of 450. Next, notice that when you match a liquid humic acid product to the quackgrass, the quackgrass vitality goes down. Ammonium sulfate (21-0-0-24S), 9-18-9, 28% nitrogen, and high-calcium lime, also lower the vitality. Notice that anhydrous, 11-52-0, and dolomite actually raise the vitality of quackgrass. Move across the page and you will get a good comparison of which products generally raise the vitality of crops and which products generally lower it. Now, compare which products lower the vitality of weeds and insects and examine what they do to crops.

Every farmer knows he can grow corn with anhydrous ammonia. This chart helps him draw conclusions as to why, after he started using anhydrous, he had to increase his toxic chemical purchases for weed and insect control plus increase his feed mineral supplement purchases. The anhydrous grows lower vitality corn while altering soil conditions to attract weeds and insects. The net result was a move toward more use of pesticides and herbicides.

The potential uses of this instrument to farming are obvious. Farmers can select seed varieties to plant in specific fields, custom program fertility to specific seed/soil combinations, and determine ration and mineral programs plus a host of other uses.

It must also be said that using an electronic scanner doesn't necessarily mean accurate results are always obtained. The scanner is operator dependent; it works well for some individuals and not so well for others. It is as effective as the operator. Manufacturers are designing units completely automated to take the "human element" out. Perhaps these will be more reliable.

Soil Test Validity

Several years ago, *The New Farm* magazine sent approximately 70 soil samples from the same soil source to different soil testing labs in the United States. In return, *The New Farm* received test results with nitrogen recommendations ranging from applying 230 pounds to 0 pounds per acre. They concluded that the concept of soil testing was not at all standardized. Upon closer examination, it became apparent that the presuppositions of the testing lab played a large role in the subsequent recommendations. For example, some labs considered estimated nitrogen release from organic matter while others did not.

Using Test Results

As the conventional soil test technology has its drawbacks due to variability in testing procedures, philosophy regarding amounts of nutrients required to grow a crop and where those nutrients should come from — from bacterial activity or exclusively from added NPK — so do the other testing approaches discussed in this chapter.

The electronic scanner technology is subject to operator variability. Of the three approaches to soil testing discussed, the scanner has the most potential for success and failure. Because of the variety of tasks it can assist in, it can be useful in unlocking a number of doors formally considered closed. At the same time, it is operator dependent, meaning that some operators will be more successful than others.

Chapter 7
Interpreting Soil Test Information

The authors have seen problems associated with using all of the previously listed types of soil testing, especially when only one was used to the exclusion of the others. With so many tools available, why take a chance at narrowing your information pool? To understand how to read a soil test and know how to interpret is to become your own fertility consultant. What do those numbers mean? What does a soil with those numbers "look" like? How do you use the numbers to bring about change?

Interpreting the CEC Test

Besides the typical identifying information consisting of name, address, lab number, and who to send the report to, the CEC test usually provides the following information:

Field ID	=	Field Identification Code
OM	=	organic matter/humus
ENR	=	estimated nitrogen release
P1	=	the "available" phosphorus
P2	=	total phosphorus

K	=	potassium
Ca	=	calcium
Mg	=	magnesium
pH	=	a water soluble reading
CEC	=	a calculated value of
		Cation Exchange Capacity
%Ca	=	percent of the total CEC
%Mg	=	percent of the total CEC
%K	=	percent of the total CEC
%Na	=	percent of the total CEC
%H	=	percent of the total CEC

If you request trace minerals they will usually be listed in a separate section of the form. The chemical symbols used mean:

B	=	boron
Ca	=	calcium
Co	=	cobalt
Cu	=	copper
Fe	=	iron
H	=	hydrogen
K	=	potassium
K_2O	=	potash
Mg	=	magnesium
Mn	=	manganese
Mo	=	molybdenum
N	=	nitrogen
NO_3	=	nitrate nitrogen
NH_4	=	ammonium nitrogen
Na	=	sodium
P	=	phosphorus
P_2O_5	=	phosphate
S	=	sulfur
SO_4	=	sulfate
Zn	=	zinc

OM: Although technically speaking organic matter is organic trash, manure, etc. in the soil prior to being broken down by soil microbes, many labs measure what we consider humus, organic matter which has been broken down and is now relatively stable in the soil. This value provides a clue as to the color of the soil, since soils with a higher humus content are typically darker in color. For example, sand can be anywhere in color from white to cream but sands with an OM of 2 percent or more will generally be characterized by a darker color. The OM reading also provides a clue as to the microbial activity. Soils higher in active beneficial microbial activity will be higher in humus. Soils lower in OM will generally have reduced microbial activity. If a lab is simply measuring total carbon material and calling it an OM reading, then it is difficult to ascertain the actual humus.

ENR: Estimated nitrogen release gives an estimate as to how much nitrogen you can expect your soil to release. Nitrogen release is directly related to the amount of OM found in the soil. Soils higher in organic matter will provide a higher amount of nitrogen. Humus (OM) does not give up it's nitrogen (and trace minerals) quickly but rather slowly over a period of years. At some point humus will no longer give up or provide nitrogen (and trace minerals) which means it must continually be generated if you wish not to purchase all your nitrogen off the farm.

Nitrogen release from organic matter depends upon a variety of factors including soil temperature, numbers and varieties of beneficial "composting" and/or nitrogen fixing microbes, the amount and quality of the food source (low brix corn stover has less energy for the microbes than that of high brix), soil air space, water, and nutrient availabilities. The reader must also remember that OM can break down into an ash (oxidize) and it can be preserved via the formation of an aldehyde (such as formaldehyde) which, as it does for laboratory specimens, preserves rather than composts. (Have you ever plowed up whole corn stocks buried for a year or more?) The authors take the stance that bacteria are negatively impacted when soil is compacted, their food source is low, chlorides are high (high amounts of potassium chloride) and magnesium levels are high. Additionally, pesticides (words ending with -*cide* mean kill), especially the use of pesticide combinations, have an unknown but presumed negative impact on most microbial life.

P1 & P2: The P1 tests for what the CEC labs call available phosphorus while the P2 tests for total phosphorus, available and reserve. These readings are recorded as P in ppm (parts per million) and the total needs to be multiplied by 4.6 to give pounds of phosphate per acre.

Historically soil labs reported phosphorus as P_2O_5 and potassium as K_2O (phosphate and potash respectively) and provided totals as pounds per acre. Many labs are now switching to record P (phosphorus) and K (potassium) in ppm (parts per million). This gives a total for P and K but not for P_2O_5 or K_2O which is what you buy as fertilizer. Multiply the P in ppm by 4.6 to get P_2O_5 (phosphate) and multiply the K in ppm by 2.4 to get K_2O (potash).

Although these calculations can be confusing, they are necessary in order to determine amounts of fertilizer to apply.

Ca: Calcium is measured in ppm. Multiply this figure by 2 for pounds per acre. Note: Since we usually consider an acre of soil to weigh 2,000,000 pounds (sand will weigh heavier while muck or peat will weigh lighter) we then multiply ppm (parts per million) by two (2) to yield a pounds per acre figure.

Mg: Magnesium is measured in ppm. Multiply this figure by 2 for pounds/A.

pH: The pH reading assists in determining the type of program you will use. pH is poorly understood by most farmers and poorly presented by most agronomists including extension agents. Although typically used to determine whether or not a field needs liming, it only indirectly provides this information and then only in relation to other data provided on the test result. A more complete presentation of pH is given in Chapter 12, but for this discussion the biggest problem with pH is that it does not provide information about whether or not to lime because it does not provide data relative to calcium, magnesium, potassium or sodium, which can all affect pH. Additionally, it is important to know which pH test your lab is using. A water pH test reads about 1 to 1.5 points higher than the salt pH test.

CEC: The CEC, or cation-exchange-capacity, is a concept which relates to the soils' nutrient holding capacity. CEC values are calculated from the figures obtained from nutrient extraction. Clay and humus are both negatively charged electrically and hold onto the positively charged cations of calcium, magnesium, potassium, sodium, some trace nutrients and some forms of nitrogen. The amounts of clay and humus determine the CEC value. The CEC value essentially tells "how big" the basket is in which you grow. By this is meant that soils with a higher CEC have more capacity to hold nutrients (bigger basket), whereas soils with a lower CEC won't hold as much. The lower the CEC, the less nutrients the soil will hold and the more attention must be paid to that soil. It's easier to make changes in the lower CEC soils because smaller amounts of nutrients are needed to make the changes, but they will need to be made more often.

Sandy soils have a low CEC and characteristically drain easily and require frequent applications of water and nutrients. The larger the CEC, the more nutrients are being held and the more difficult it is to make changes in the nutrient ratios because of the need for applying larger amounts of nutrients. Clay soils are heavy, tend to hold moisture longer, take longer to dry out, and are capable of growing a crop often with little or no added fertilizer except, perhaps, for some starter. Knowing the CEC values for soils, e.g., sandy (0-4), loamy (5-8), clay (11-15) and muck or peat (20 and up), provides you with an immediate conceptualization of the type of soil identified by the soil test.

Percent Base Saturation: The percentage of nutrients — %Ca, %Mg, %K, %Na, and %H — relate directly to the pounds measured. This is a calculated value arrived at by dividing the pounds per acre of each nutrient by the total weight of all nutrients measured. Because soil holding capacities are directly related to CEC values, ideal nutrient levels are generally given in nutrient ratios as well as in pounds. Although we may indicate these ideal ratios, CEC levels themselves determine how quickly nutrient changes can be made in the field.

"Seeing" the Soil Through CEC

With the above in mind you can now develop your skills to see, although perhaps in a general way, what a soil looks like just from reading the CEC soil test. Higher organic matter levels indicate a darker soil, one more rich in microbial life. The higher the microbial life the more available the nutrients will be since one of the jobs microbes fulfill is to make nutrients available for plant life.

In the traditional Reams model the Ca:Mg ratios would be 7:1 (some say 10:1) to indicate a well-balanced soil. Narrower Ca:Mg ratios, say, 4:1, indicate compaction. The tighter the soil the less drainage and the less favorable the soil for microbes. This means the soil will be harder to till requiring more horsepower and heavier equipment. Sour grasses such as foxtail, crabgrass, etc., can be expected in this environment. (See *Weeds and Why They Grow* by Jay McCaman for a more complete understanding of what weeds tell about the soil.) The less microbial action the less calcium will be available which is the explanation for their presence — foxtail and quackgrass help make calcium available.

According to Albrecht, the ideal soil would have the following levels: Ca – 68%; Mg – 12%; K – 2-5%, and H – 10-15% with a pH of 6.2 to 6.5. Neal Kinsey, a disciple of Dr. Albrecht, has modified this ideal when CEC's fall below 7. In these cases he would lower the Ca to 60 percent and raise the Mg to 20 percent. This change would provide for additional magnesium which would help hold the soil together while at the same time reduce the Ca which, by definition, helps loosen the soil. These recommendations are based upon the original Albrecht CEC testing procedures.

LaMotte Test

The soil test sold by the LaMotte Chemical Company was modified and used by Reams to measure what he called *available plant nutrients* — those nutrients in the soil solution immediately available to the growing plant. Because nutrients become available for the growing plant from root extracts produced by the plant itself, microbial activity, and water-soluble nutrients applied as fertilizers, Reams felt it important to monitor which nutrients are or are not available at any particular time. Since available nutrients may fluctuate from day to day based upon rainfall, crop demands, etc., frequent monitoring helps prevent yield loss by indicating nutrient deficiencies before they show up in the plant. Reams promoted the purchase of LaMotte soil testing kits and recommended growers test their soils at least twice a month. The LaMotte test is usually recorded in pounds per acre and provides the following information:

$$
\begin{aligned}
AN &= \text{ammonia nitrogen} \\
NN &= \text{nitrate nitrogen} \\
P_2O_5 &= \text{phosphate} \\
K_2O &= \text{potash} \\
Ca &= \text{calcium} \\
Mg &= \text{magnesium} \\
SO_4 &= \text{sulfate} \\
pH &= \text{a water soluble reading} \\
Na &= \text{sodium}
\end{aligned}
$$

ERGS — Energy released per gram per second, or conductivity reading.

ORP — Oxidation Reduction Potential, a value indicating whether the soil is composting (taking up) or releasing (giving off) energy.

Interpreting the LaMotte Readings

AN & NN. As will be discussed more completely in Chapter 11, nitrogen isn't "nitrogen." Plants utilize two forms of non-organic nitrogen — nitrate, which leaches and promotes growth, and ammonia which doesn't leach and promotes fruiting. Both are needed by the plant and soil environment. As already discussed, soils low in microbial life and humus hold relatively small amounts of ammonia nitrogen which relates directly to insect pressures. Nitrogen is also taken up in organic forms, usually as parts of or whole amino acids.

Since nitrogen can volatilize depending upon how wet the soil sample is and how long the sample sets before being tested, it's a good idea to fan dry the soil overnight before mailing to a CEC lab, while some LaMotte labs prefer to test soil as it comes from the field.

ORP readings are obtained using the ORP meter and calculating the result based upon the soil's pH. ORP readings indicate whether your soil is oxidizing (aerobic decomposition) or reducing (composting). Since oxygen is the key to the composting operation, it's important to be able to measure what is happening and respond accordingly.

The ideal ORP range is between 25 and 29. Soils with a reading lower than 20 can be said to be greatly lacking oxygen due to its use in the composting process. Such soils are characterized as poor growth mediums. Seeds planted in these soils may tend to rot as there will be an excess of moisture. Soils between 20 and 25 are also lacking oxygen. Decay in these soils is characterized by anaerobic decomposition with the resultant production of toxins such as formaldehyde. This is a condition to be avoided since most beneficial microbial life favor aerobic soil conditions.

Soils higher than 32 are strongly oxidizing (burning out) and are rapidly releasing energy and loosing humus. These soils provide too dry a medium and may prove a difficult medium in which to germinate seeds. In composting operations, readings higher than 32 will have an excess of carbon dioxide which will be lost to the atmosphere — essentially the compost is burning itself up. Basically the ORP meter is measuring oxygen or lack thereof. Oxygen is the driving force for all of nature.

Na. Sodium readings are typically high in western soils, but generally Na is not a factor in midwest or eastern soils except in certain situations. These include soils treated with high amounts of manure or compost and in areas of the country where salt spills or naturally occurring salt mines are present. When ERGS are high, e.g. 800 or higher, check to see if sodium is a factor. High-sodium soils can be more difficult to grow on. Very low sodium soils can benefit from applications of sodium for flavor enhancement of produce and soil texture.

Nutrient ratios are important to evaluate when using the Reams test. The Ca:Mg ratio 7:1 is preferred. Ratios less than this will generally indicate sour grass pressures. The P:K ratio wants to be 2:1. Ratios less than this will generally indicate broadleaf weed pressures. The K:S ratio should be about 1:1.

Field Meters

As part of field observations, it would be helpful to be able to take some on-site readings. There are a variety of new portable meters in the marketplace.

Six portable instruments deserve mention for farmer use: refractometer, pH, ERGS, sodium, and ORP meters, and the new magnetic susceptibility meter developed by Dr. Philip Callahan. These are affordable, can be carried in the pocket or in a small carrying case, and can give immediate on-site readings. The refractometer will be discussed in detail in Chapter 14 and is used for measuring plant juices which gives an indication as to the health of the plant/soil combination. The remaining meters are used to measure soil solutions.

The portable, inexpensive ERGS, pH, Na and ORP meters have a life expectancy of about two years and give dependable readings. They are battery operated and provide a digital readout. The paramagnetism meter is a larger bench meter and should have years of useful life.

pH Meter

The need for and use of pH readings is covered in another chapter and will not be covered here. However, a pH reading is required to interpret the ORP reading. The ORP meter will be discussed later in this chapter.

ERGS Meter

One of the most important readings requires an ERGS meter. ERGS means Energy Released per Gram per Second in the Reams' system; or more conventionally, it is known as a soluble salts or conductivity meter reading 10 to 1,990 microSiemens (μS). This meter gives readings of the amount of both negative and positive ions that are flowing in the soil. This would be an indication of the amount of nutrients (energy) available to the plant at that particular time. The readings will change quite rapidly if soluble fertilizers are added to a soil, or the change can occur more gradually due to a rain, irrigation, or drying conditions which changes the concentration of the soluble ions. Dr. Reams taught that the best range for growing crops was 200 to 400 ERGS over base. Base was defined as an ERGS reading taken from a fence line or an unfarmed portion of the farm that was not receiving fertilizer. For example, he would subtract a base reading of 100 from a field reading of 300 resulting in a reading of 200 ERGS. This number would be adequate for early spring soils, but would be too low for a growing crop, especially in later development stages as found during the month of August.

Users of his technology are finding that although the base reading concept is still valid, the ranges of ERGS one has to work with may be much higher than the 200-400 range. Part of the reason may be that the plant hybridization process has developed plants needing to consume or grow in the presence of high amounts of nitrogen which can result in higher ion counts. However, although plants may grow at the higher ERGS levels, the bacterial populations may not function well enough to result in high brix readings along with the potentially higher production. Some of Dr. Wheeler's clients are running as high as 1,200-2,000 ERGS under plastic with drip irrigation and achieving excellent results.

Sodium Meter

Sodium may also be a factor in the higher ERGS readings which brings the pNa or sodium ion concentration meter into play. One can verify whether sodium is a substantial part of the ERGS by using this meter. Sodium meters are especially important for Western growers. The lack of sodium seems to be a greater concern in Eastern and Midwestern soils. Sodium can be an important part of the CEC complex and a total absence affects the flavor of fruits and vegetables.

ORP Meter

The newest concept to be used for soil is the ORP measurement which stands for Oxidation Reduction Potential. The reading is used to evaluate the basic soil process of oxidizing humus for release and the use of this energy by crops and soil life forms. The ORP reading is more familiar in the compost industry where it is used to determine compost stability. It also could be used to evaluate the ensiling (reduction) process. As stated above, a pH reading is required along with the ORP reading to deliver the rH value. This value is determined from a chart supplied with the meter.

The ORP concept has been investigated extensively in Europe. Basically, it tells whether or not the humus is working in the soil.

The use of these meters provides the grower with hands-on tools and allows for on-site analysis of soil conditions on a daily basis. This information should closely parallel LaMotte soil test results thereby giving the grower fairly accurate information as to what is happening under his crop. Operation is similar for the pH, ERGS, Na and ORP meters. Simply take a given measure of soil and mix with an equal measure of distilled water. Stir gently for about 30 seconds and allow to settle for an equal length of time. Take the protective cap off the meter end and place the sensing bulb into the soil solution. Shake gently to expel any air bubbles from around the bulb. You may need to wait for 15 seconds or longer for the reading to stabilize. Once the reading is taken, you need to clean the bulb off with distilled water.

In some cases the bulb will need to be kept moist and/or the meter will need to be calibrated. The supply house that sells you the meters will have the appropriate solutions for each purpose.

Although the information provided by each meter may be limited and, by definition, incomplete, each does provide information from which to make program adjustments. Obviously more complete information will require the taking of CEC and/or LaMotte soil tests.

Paramagnetism Meter

The paramagnetism meter measures the ability of the soil to tune into and receive magnetic energies of the cosmos. The meter measures this ability electronically. According to physicists, all matter has an electrical ability to either be attracted (paramagnetism) or repelled from (diamagnetism) a magnet. Plants are diamagnetic. Air, water and soil are paramagnetic. Stone powders (see *Paramagnetism* by Dr. Phil Callahan and *The Enlivened Rock Powders* by Harvey Lisle) are mostly paramagnetic, but the extent of their paramagnetism must be measured. The more highly paramagnetic, the better the response of the growing plant. In the classic book, *Bread from Stones,* Julius Hensel recaps how crops were successfully grown by only applying rock powders. The secret here is the extent the powders were paramagnetic. Paramagnetic soil qualities are "anchored" in place via the use of composts or any other method of producing biologically active carbon. Using the paramagnetism meter the grower can determine how well his soil is able to use the free energy nature provides. Basically the following readings (units of attractive force) tell the answer:

0-100	=	not good soil
100-300	=	good soil
300-700	=	very good soil
700-1,200	=	excellent soil

The authors tested some Brazilian red clay at 1,200 units. The owner reported being able to just plant and add water to produce good vegetables.

In personal telephone conversations with Dr. Callahan, we learned that he tested all of the insecticides "known to man" working with LSU. He found them all to be toxic to the soil. Standard chemical farming does not take into consideration the paramagnetic qualities of the soil or anything applied to the soil. Is it any wonder that nutritional testing shows crops grown on such soils to be deficient in vitamin and mineral levels while, at the same time, insect pressures continue to rise? According to

Callahan, "Every sacred place I've been to is highly paramagnetic." Sacred places leave the person feeling energized and full of vitality. So too for the crop.

The question might be posed as how to increase the paramagnetic energies of your soil. We think it is easy to explain its destruction compared to trying to identify a set process for restoring it. Application of highly paramagnetic rock powders is the most direct way; the authors have seen both good results and no results from this method. Restoring is more complex and may occur in response to a variety of additives or cultural practices. When the meter detects it, it's back.

Interpreting the Scanner Test

The radionic scanner test is totally different from either of the above soil tests. Rather than providing "pounds per acre" data, it provides "energy intensity" readings which are compared with the general vitality (GV) reading.

Most scanner labs provide the following information in their standard lab reports:

GV — general vitality
Nitrate
Ammonia
Phosphate
Potash
Calcium
Magnesium
Boron
Chlorine
Copper
Iron
Manganese
Selenium
Sulfate
Zinc
Brix
Carbon
pH

Aluminum**
Chemical toxins**
Metal toxins**
Salt**
Pathogens**
** = should read below 50

GV. General vitality can be considered as the basic energy level of the plant. Improving the GV, if everything else is proportional, improves the health of the plant. Reducing the GV does the opposite.

To interpret the scanner test one must compare the given nutrient energy readings to the general vitality. The extent the nutrient reading varies from the GV, to that extent it is in excess or deficient. Generally speaking, a variance of 50 points or 10 percent is not considered crucial depending upon the GV reading itself. Notice that the scanner also measures those factors which are considered harmful or not desirable. Salt and chemical or metal toxicities have a harmful effect on soil life and plant life. It they accumulate because of the inputs you are using or due to your management approach, then it can be said that you are contributing to your own problem.

Being able to identify problems also brings the possibility of correcting them. Since the scanner is measuring energy, these readings indicate which nutrients are providing the specific energy pattern in the soil, or, if measuring the plant itself, which have entered the plant and are providing the specific component there. Plants low in a particular nutrient, as measured by the scanner, may or may not benefit from the addition of that particular nutrient. As stated elsewhere in this handbook, many times nutrients will tie-up and/or otherwise be unavailable to the system, e.g. plant, soil microbes, etc. Soils testing low in calcium may already have sufficient amounts of calcium present, albeit in a form which the plant can't use. It is in situations like this that the scanner, in the hands of a competent operator, is most helpful.

Rather than using the scanner to test soil as per the above values, most labs use the scanner to help determine which nutrients to apply to a soil. By measuring the GV of a soil and then adding the "energy field" of a certain product, the operator can determine if the product being measured will raise, lower or leave untouched the soil or plant GV. Additionally, the scanner can be used to help determine which seed varieties or plant varieties will raise the soil GV the most. In other instances the operator will help determine which field raises the seed GV the most. Planting decisions can be made accordingly. Once a crop is out of the ground, the scanner can help the grower determine which foliar or sidedress programs will raise the composite soil-plant GV the most. Just as your body's nutrient needs vary over time, so does the crop's.

Factors such as stress, heat, cold, insect pressure, a cold or hot summer, etc. all contribute to the necessity of evaluating the crop on a periodic basis. For example, if the crop shows a calcium deficiency it may be that adding vitamin B12 to the soil may encourage the microbial activity to break loose calcium making it available to the crop. Conventional agriculture would never have considered an application of vitamin B12, yet this may very well be the key to unlocking the system.

After soil testing, the next important thing to do is to observe what you look at and to document activity. Recording weekly portable test results, including refractometer readings, can give an excellent database from which to begin learning how the system works and how it responds to specific inputs. As this skill is enhanced, growing becomes easier and you'll have more time to respond to other demands.

Obviously the farmer can't test on a daily basis and can't apply nutrients on a daily basis either. Yet all the tests taken provide data from which to make the necessary management decisions at whatever frequency is appropriate to the crop and its relative value.

Chapter 8

Making Fertility Recommendations

The first step in utilizing the test information would be in evaluating the CEC test. How high is the CEC (how big is your container)? What are the percent saturations of Ca, Mg, K, Na, traces and H? If the percentages are low, materials will need to be applied, especially when it comes to Ca and K. In some cases, Mg may need to be applied, especially in low-CEC soils.

Calcium was considered by both Reams and Albrecht to be of prime importance. Ca is directly related to microbial activity, soil flocculation, soil drainage, and overall soil performance.

Magnesium is often found in excess and considered a problem. This is true when in excess of 15 percent base saturation. In sandy low-CEC soils, however, magnesium may be needed up to 20 percent both to supply a minimum of 200 pounds per acre and also to provide soil holding power.

Potassium is also important to consider. The authors have heard many a farmer report that by following biological programs their potassium levels increased while no potassium was added. At the same time, if potassium is short, it must be added. This is particularly important when available calcium levels are low.

Compared to the CEC, test readings on the LaMotte test generally will be lower, in some cases significantly lower. Since the LaMotte test is believed to test nutrients available to the plant, the differences observed between the two tests are indicative of the microbial action taking place in the soil. If, for example, the CEC test shows high levels of calcium and the LaMotte shows low levels, it is understood that the calcium is present in the soil but tied up and in forms not available to the plant.

You may find that some nutrient levels are higher on the LaMotte test than the CEC. In this case it is presumed that good microbial action is making the nutrients available to plants or that available forms have been applied. There may not be much in reserve in this situation, so it may be necessary to add the nutrients that tested low on the CEC. After receiving the test results, the farmer must now make economic and management decisions and put them into effect. To help complicate matters, a soil calling for 400 pounds of a given nutrient, but with only 200 pounds applied, may retest the following year calling for only 100 pounds additional. This is especially true when the biological life is active.

Using the CEC and LaMotte Tests

It's important to compare several variables both within and between the two tests in developing fertility programs. First, compare the ratios as discussed above. Are they similar? Do they indicate weed pressures? What type of weeds? Can you see evidence of compaction, or improper decay?

Second, compare the OM level and nutrient levels. Do these give an indication that microbial life is working or not? What type of N levels can you expect under good conditions? Compare CEC levels to LaMotte levels. Are nutrients missing from the soil or are they present in sufficient levels but need to be activated? If nutrient levels are low, you need to know the nutrient forms which will provide the needed nutrient plus provide the best impact on microbial life (see Chapter 10).

Consider the ERGS reading. Low ERGS means low nutrient support. Which nutrient is short? The ERGS reading doesn't tell you that information. You must have the complete LaMotte test to determine this. Using the portable ERGS meter is still an excellent way, however, to monitor a soil's performance throughout the year.

Consider the humus level. When OM (humus) or microbial levels are low, sustainable farming is difficult to achieve, especially when you desire a reduction of off-site inputs.

Microbial life is essential to non-toxic, sustainable or biological farming, and an analysis must be made as to the cause of its decline. Are fertilizers damaging to microbes being utilized? Are low-brix/low-energy crops being grown and plowed down to feed the microbes? Are toxic chemicals being used which may play a role? Is the soil tight from excess magnesium and/or lacking in calcium?

When nutrients are low on a CEC test they usually need to be added. When the calcium percentage shows less than 60 percent, many (most?) microbial products, including humic acid, don't work that well. Microbes need calcium to live. At the same time, too high magnesium levels are also damaging to microbes. If calcium is low, add high-calcium lime. Ask for an analysis of the lime prior to purchase. Some quarries have high-cal lime which is about 35 percent Ca and 5 percent Mg, a 7:1 ratio.

When calcium levels are adequate (at least 60 percent) and the LaMotte Ca test reading is low, the calcium in the soil needs to be activated. This can be done by adding a liquid calcium product. One LaMotte lab owner has found that 2 gallons of liquid calcium raised the available calcium level on the LaMotte test by 2,000 pounds per acre. This indicates the ability of this type of product to release nutrients. Another approach would be to add compost or manure to stimulate microbial action on the calcium.

According to soil fertility consultant Neal Kinsey, soils less than 60 percent calcium will not be open enough to leach what may be in excess. For example, if Mg is in excess, the Ca level must be 60 percent or higher if sulfur is to combine with the Mg (forming epsom salts) and leach out. Low available calcium levels will show up on the refractometer. The line at which the blue and white fields join is the indicator. If the line is real sharp, a calcium deficiency is indicated. If the line is fuzzy, a good supply of calcium is indicated. This is based on the assumption that calcium causes a more alkaline pH in the plant juices.

Potassium is one nutrient which is often over applied. Conventional agriculture has emphasized nitrogen and potassium. Crops grown under this program tend to look nice and green, but also tend to be higher in nitrates, lower in sugars and often tend to lodge more easily. We call alfalfa grown like this "gunpowder hay" because of the nitrate and potassium (potassium nitrate or saltpeter) levels. Biological farmers often report an increase in LaMotte results for potassium even though no potassium had been applied for several years. Since total content of potassium in soils is often very high, microbial stimulation can make potash available. At the same time, if potassium levels are still too low for current crop production, potassium needs to be applied. When phosphate levels are low you can expect lower refractometer readings. Plants need phosphate to make sugars.

When ERGS levels are low, e.g., less than 100, plant growth will be slow. What has caused this? Has rainfall been excessive? Is the soil low in water-soluble nutrients? If the former, testing after the soil dries out will see an increase in ERGS. If the latter, you may need to side dress or foliar apply fertilizer.

Obviously the first step in applying materials is to stop using the bad guys, those materials which have a damaging effect on microbes. Mother Nature is very forgiving and, if we will do our part by reducing or eliminating those inputs which negatively effect the soil, she will do her part in rejuvenation.

Equally obvious is the need to enhance microbial action. Although arguments can be made against the use of microbial products (if the soil can't hold them why inoculate with them?), other steps can provide an increase in beneficial microbial levels. The use of composts is one of the best examples. Reams recommended the use of poultry manure because of its microbial content. Other manures can be beneficial, depending upon the condition of the manure. If possible, compost the manure before using it.

The use of carbohydrates on the soil such as molasses and various kinds of sugars can improve the food source of microbes. As you begin to build fertility levels and see the benefits via improved crops, the stubble you return will further increase microbial life because of its increased brix or nutrition. This is an endless spiral. Direct it in one direction and you see increasing soil damage and need for off-site inputs. Direct it the other direction and soil improvements are witnessed, plus a decreased need for inputs as nutrient availability increases.

Many "biological" or microbe-containing products are available in the marketplace. Used with judicious care they can greatly enhance recovery operations. Of all the comments the authors have heard from farmers using the Reams approach, none have ever thought they needed less biological activity.

In many Reams circles you hear that you can never apply too much calcium. This is because of the emphasis he placed on the role of calcium in the growing of plants as well as in the workings of the soil. This advice, however, may not be completely correct. Applying lime, by definition, means that other nutrients may be limited in the process. Calcium, for instance, can reduce potash availability. Calcium can also tie up or keep plants from taking up trace minerals such as boron. According to *Hands-On Agronomy*, excess calcium can hide magnesium. If Neal Kinsey is correct in this respect, too much calcium fools the reader by concluding that magnesium is in the correct range whereas it actually is in excess.

When any cation — Ca, Mg, K or Na — is in excess, the standard solution is to apply sulfur, usually in the form of sulfate. In these situations, sulfate ions will associate with the cations in a loosely held soluble form and leach downward with rain or irrigation water. A combination of sulfur with the magnesium results in a leachable compound called epsom salts. Gypsum, ammonium sulfate or ammonium thiosulfate can all be used in this situation. Be careful not to apply much more than three gallons thiosul per acre, however, as negative side effects can result when an excess is applied.

Farmers have another option when potassium levels are high. Cropping of potassium-loving plants, such as alfalfa, removes the K in the harvested crop and it can be sold off the farm. Current attention to phosphate and nitrogen levels may result in the legislative restriction of phosphate and nitrogen fertilizers. Phosphate, as all minerals, needs microbial activity to make it available to the growing crop. Although phosphate may test high on a CEC test, it's important to have a LaMotte test to verify whether or not it actually is available. Phosphate, like calcium, often tests low on the LaMotte as compared to the CEC test. Improving microbial life will also improve available phosphate if it is present in reserve or tied-up form.

Most soil nutrients can be made available when small amounts are applied in an "available" form. This is especially true when used in liquid form. Liquid calcium is excellent in making soil calcium more available and is often recommended at 2 gallons per acre with equal amounts of blackstrap molasses for phosphorus release. This combination can also be used as a temporary weed and grass suppressant. These programs may be just what is needed to help turn your soil around. The availability approach may be more appropriate than CEC balancing on high CEC soils because of costs. That's why the authors recommend the use of the three testing procedures to determine the best approach.

To better understand why testers using the electronic scanner often recommend fertilizers other than the typically applied products, it is important to understand what is problematic with commonly used fertilizers. This is the topic of the next chapter.

Chapter 9
What's Wrong
with Today's Fertilizers?

Background

Chemical fertilizers have provided increased yields, bolstered farm income, and brought marginal land into productivity. What could possibly be wrong with them?

Fifty years ago, fertilizer was applied at the rate of 50 pounds per acre. Now, applications of 500-800 pounds are common. What has caused this increased dependence upon chemical fertilizers? What other problems, aside from economic requirements, are they responsible for and what alternatives exist?

When fertilizers other than manures first came into use in ancient times, they consisted of rock minerals, such as limestone, mined from the earth and applied directly to the soil. These provided minerals which were lacking in the soil. Rock minerals were usually applied in addition to manures and were considered supplementary in nature.

Today, many farmers do not apply manures and consider fertilizers to be primary. The rock minerals have given way to

commercially-processed "chemical" fertilizers. The triple num-
bered dry fertilizers, e.g., 16-16-16, contain nitrogen, phosphate,
and potash, the supposed "main nutrients" a plant needs. This
simplistic notion of fertilization is what has caused many of
today's agricultural problems in spite of increased yields, problems such as crops which are eaten by insects, crops which are
refused by other countries due to their low mineral (specific gravity) content and erosion at critical levels caused by soil, wind, and
water degradation.

Problems

The first reason for this trouble is that the NPK concept is
actually incorrect. In the early 1800s, a German scientist, Justus
von Leibig, concluded after burning and analyzing plant issue
that three basic nutrients, nitrogen-phosphorus-potash (NPK)
were needed to grow a plant. His conclusions were picked up by
the universities and business communities of the time. Later in
his life, he acknowledged that his initial conclusions were in
error. He realized that more than these three elements were needed for proper plant growth. What was also needed was a supply
of trace minerals, microbial life, and humus. This update was
never acknowledged. The "system" was moving and preferred,
instead, to remain with the simplistic NPK approach. Dr. von
Leibig was correct in stating that plants need a much greater variety of nutrient base to grow from than three basic nutrients.
Plants require trace minerals and so-called "secondary" nutrients
the same as animals do, and this fact has only recently been generally recognized. Calcium, for instance, is now being recognized
as crucial to the growth and health of plants.

Other supposed "secondary" nutrients, such as sulfur and
magnesium, also play important roles as well as the "trace" nutrients, e.g., iron, manganese, and copper. The current proliferation
of trace mineral additives on the market speaks to this realization.
By ignoring the need for a "complete diet," today's farmer has
greatly depleted his land of the needed trace and secondary nutrients. The resulting poor health of his soil and crops are his evi-

dence. Poor refractometer (an instrument which measures the "sugars" or dissolved solids content in plants) readings indicate poor nutrient or mineral levels within the crop itself. As these nutrients are replaced in the soil and assimilated by the crop, the refractometer readings increase dramatically and insect attack reduces dramatically as well.

A second reason for the failure of the simplistic NPK concept is the ingredient makeup going into the early NPK fertilizers. For instance, the nitrogen portion typically was urea (and often still is). The urea tended to have significant formaldehyde residues and was subject to volatilization if not incorporated. The P used in typical dry fertilizers was triple superphosphate (0-46-0). In biologically healthy soil, triple superphosphate is only 15-20 percent available to the plant. A biologically healthy soil means a soil with abundant, active microbial life. In biologically dead soils the availability would be less. The 0-46-0 readily ties up with calcium in the soil forming tricalcium phosphate which means that the calcium and the phosphate are now both unavailable to the plant. Not only is the product extremely expensive (15 percent or less availability is expensive), but tying up calcium causes serious problems as will be discussed later.

The third ingredient still typically used is the most commonly used form of potash. The potassium source is usually a form of muriate of potash (KCl 0-0-60). Muriate of potash contains approximately 50 percent potassium and 50 percent chloride. The potassium chloride can be compared to table salt which is sodium chloride. Both taste like "salt," but one contains sodium while the other contains potassium combined with a chloride ion.

Salts

The effect of chloride on the soil is not good. The immediate negative effect of small quantities is probably very difficult to measure, but increasing the amounts used accelerates the damage. The chloride ions kill off the prevailing "good bugs" (microbes) within the soil which are necessary to soil rejuvenation and crop productivity. It is these microbes which break down organic matter into humus and which actually consume the fertilizer applied by the farmer, hold it in the soil, and make it available to the plant.

By applying small amounts of "chloride-salt" fertilizer at one time, the farmer has not been able to recognize the gradual damage being caused. Now, with increasing amounts being applied, the farmer wonders why he continually plows up last year's corn, why increasingly larger amounts of fertilizer are required, why he has increasing weed pressures, and why the organic and humus levels in his soil are bottoming out, causing greater soil erosion today than was present in the dust bowl days of the 1940s. This has occurred despite 50 years of research and advice from the USDA Soil Conservation Service.

The fact is that chlorine gas and chloride ions are both highly effective in killing bacteria and other soil microbes. Chlorine is used in municipal water treatment plants to kill bacteria in city water. One hundred pounds of muriate of potash per acre has been compared to 100 gallons of Clorox being distributed over an acre of soil. In fact, when chlorides combine with nitrates, chlorine is released. Through continued use of this soil "killer," the desired aerobic microbial life has been seriously depleted and/or changed in character. Compaction has induced the anaerobic bacteria supposedly found only in the lower levels of the soil to populate the majority of the soil bed. Potassium chloride isn't the only culprit. Herbicides, pesticides, and other farm chemicals also contribute to the decrease of proper soil life.

This destruction of beneficial soil populations means incomplete or improper plant decay. For proper soil tilth, organic matter must be decomposed into humus. In typical conditions of improper decay, an aldehyde is formed. It is not unusual, using the electronic scanner, to find significant levels of formaldehyde present in the soil. This preserver is known for its embalming qualities. Thus, the soil increasingly becomes an embalming medium, proper decay no longer occurs, and humus levels decline.

Once the soil has been damaged in this manner, plants require increased amounts of readily available plant food, such as obtained from soluble fertilizers, to survive. These are subject to leaching from the soil with resulting ground water contamination. One soil and crop consultant has stated that "potassium chloride is the number one cause of today's farming problems."

It must be stated that potassium chloride is not the only fertilizer with a "salt" index. All dry chemical fertilizers have a salt index. Organic farming adherents have grouped all the chemical fertilizers together because of their salt indexes and, therefore, have refused to use any of them. The authors treat them individually based upon testing results which show some to positively benefit plant and soil life. For our discussion, "salt" fertilizers mean those blended fertilizers containing potassium chloride.

The third reason for the problems associated with fertilizers is an elaboration of number two above. Aerobic bacteria are those bacteria which are oxygen loving and therefore live in the top several inches of topsoil. Topsoil is generally considered the aerobic bacteria soil layer. This soil layer is characterized by high microbe activity, looseness, and correspondingly high air content where most of the plant decomposition takes place. Excellent topsoil composition is considered to be 5 percent humus, 45 percent mineral, 25 percent water, and 25 percent air. This combination is excellent for growing plants and is fairly easy to plow and till.

Microorganism Destruction

In recent years, the topsoil layer has been greatly reduced. This is not all a by-product of chemical fertilizer, but is related to bacterial action. If bacterial numbers are reduced, they will take up less "space" and no longer work to flocculate or loosen the soil. As their numbers continue to fall, the topsoil layer will continue to degrade, and compaction will increase.

Aerobic bacteria require air (oxygen) and also calcium. Calcium, as will be discussed later, is found in lime. As agronomic advice has stressed the use of potash, it has tended to neglect calcium levels in the soil, especially if the soil pH was in acceptable levels. Clay soils require calcium in order to naturally flocculate or loosen. As increasing amounts of potassium in the form of potash is applied, the potassium occupies the place intended for calcium on the clay colloid. The clay colloid, devoid of the proper amount of calcium, then begins to "flatten out," taking up less space, contributing to soil compaction and eliminating air spaces. This same result may occur with the addition of dolomite lime and/or excessive magnesium. The soil pH rises, but no one suspects a deficiency in calcium. Not only does calcium reverse this process, but calcium is a major nutrient required to produce a response by aerobic bacteria.

The lack of aerobic bacteria also results in a reduction in soil humus content which means even less material to loosen the soil. Additionally, since soil humus is the primary reservoir of soil moisture, reducing humus levels means a great reduction in the soil's ability to hold water. Run off, drought stress, and erosion are all natural consequences.

As soils become increasingly compacted, larger and larger equipment is required to pull the same or smaller plow. As equipment becomes larger, it adds to the compaction continuing the destructive compaction cycle. Therefore, compaction is not strictly a by-product of chemical fertilizers, yet it is greatly affected by fertility programs.

The authors generally consider the use of herbicides and pesticides to have a negative impact on soil and soil life. By stressing imbalanced infertility inputs, the system (USDA, land grant universities, and conventional agriculture consultants and businesses) has produced the problems associated with increased weed pressures. These same soil imbalances result in imbalanced (unhealthy) crops which attract insects — thus the need for the chemicals to *rescue* the farmer from the conditions he has unknowingly created. The continued use of pesticides further damages the soil and crops grown on the soil and provides a potential source of contamination to the livestock and people who consume the crops.

Chapter 10

Blended Fertilizers

Fertilizers

Fertilizers are sold by grade. Fertilizer grades indicate the amount of NPK (nitrogen (N), available phosphate (P_2O_5), and soluble potash (K_2O) guaranteed by the manufacturer to be contained in the product. These numbers indicate the number of units of NPK as a percentage of 100 pounds. For instance, 100 pounds of 12-50-0 would contain 12 pounds nitrogen, 50 pounds phosphoric acid, and 0 pounds potash. Since all fertilizer is sold according to these guidelines, it is supposedly easy for the farmer to order and know what he is receiving. We say *supposedly* because fertilizers may differ radically in the amount of energy they contain due to the raw materials used in their formulation and their contaminant levels. It is impossible to know what you are buying unless you know the material source or test the product using a device such as an electronic scanner.

As was stated earlier, plants grow from energy and not from fertilizer. Fertilizers contain the energy patterns inherent in the raw materials from which they were derived. This concept allows the farmer to understand why one fertilizer will work well in a certain field while another brand, even of the same grade, will not.

Dry Fertilizers

Dry fertilizers are relatively easy to use and handle. They can be applied broadcast or in-row and have been the major form of fertilizer sold. They are made from ingredients listed in the last chapter and can be custom mixed or are sold by formulations determined by the company. Each supplier has the proverbial 101 ways to make the fertilizers they sell. By combining different sources of nitrogen, phosphorus, potash and filler, the correct fertilizer grade can be made.

The most commonly used sources for dry fertilizers are urea (46-0-0), triple superphosphate (0-46-0), and potassium chloride (0-0-60). These are relatively inexpensive to mine or manufacture which keeps the price down to the farmer. Farmers can have their fertilizers custom blended by many elevators which means they set the NPK specifications, and the elevator makes a batch of product per this specification. Unless the farmer specifically states which raw ingredients to use or not to use, however, he is still not sure what the product he is purchasing contains. Custom blends using ammonium nitrate, ammonium sulfate, sul-po-mag, potassium sulfate, 11-52-0, and other "better quality" raw ingredients will eliminate the negative soil effects discussed in Chapter 9, give good energy response, and benefit the crop and soil microbes. The farmer's main job is to know his fertilizers and require guarantees of what he is purchasing. In other words, buy what you want, not what someone has to sell.

We categorize many dry fertilizers as being "salty" and "acidic" because of how they are made. Triple superphosphate, 0-46-0, is made by treating hard rock phosphate with sulfuric and/or phosphoric acid. The result is a product having a higher phosphoric content than the original rock. It is also very acidic. Potassium chloride, 0-0-60, is composed of roughly 50 percent chloride. It is properly called a potassium salt. Potassium chloride is sold as a salt substitute for persons on a low sodium diet.

The main problems with dry acidulated fertilizers are: they usually are salty and acidic and will rust equipment; they are usually applied only once in the spring and consequently, the fertilizer energies give out when the crop needs them the most during the later growing stages; often are harmful to soil life; and they cause soil texture problems such as compaction and loss of carbon.

There are many dry fertilizers that do not match this description. They can be bio-enhancing, help to balance CECs, and provide the right energy at the right time. Dr. Wheeler makes frequent use of ammonium nitrate, ammonium sulfate, sul-po-mag, 11-52-0, and others to start a soil on the road to recovery.

Liquid Fertilizers

Liquid fertilizer blends, such as 9-18-9, 3-18-18, and 10-20-10 are a bit more difficult to produce than dry fertilizers. Mixing tanks are required, and different raw materials are used which also require different mixing procedures.

"Cold process" liquids usually are made from the least expensive raw ingredients including urea and potassium chloride. These ingredients are merely combined together and mixed with water. They usually cost the least to make as well as to buy, and they often contain the least amount of energy.

"Hot mix" liquids usually are made of more costly ingredients including feed-grade urea, phosphoric acid, and potassium hydroxide. The latter two ingredients create great heat when combined (a chemical reaction takes place when an acid and an alkali combine resulting in an entirely new chemical) thus, the name, "hot mix."

The resulting product is usually adjusted to have a near neutral pH. If the manufacturer wants to cut his costs, he can use "spent" acid which is a by-product of washing metals. Spent acid will work in making "hot mix" fertilizers; however, it will also contain large amounts of heavy metal contaminants gathered during the metal washing process. Contaminants greatly reduce the energy level of the fertilizer plus add their own negative energy pattern to the soil and crop. Clean liquids made from high-grade raw materials usually have good energy readings.

Liquid fertilizers have several advantages over dry "salt" fertilizers. First, they usually have a neutral pH which means they will not rust your equipment. Second, they are versatile in their application being used in-row on planters, broadcast from boom sprayers, and foliar applied by boom, mist blowers, or aircraft. They are easily stored and can be kept from year to year. Third, they are usually produced in beneficial phosphate-potash ratios, e.g., 9-18-9. Of course, if your machinery is set up only for dry fertilizers these advantages won't apply to you.

Generally speaking, good, high-quality liquid fertilizers usually contain more energy and are used in lesser amounts than dry fertilizers to produce the same starter effects. Since plants grow from energy, the amount of product applied can be reduced. Also, quality liquids often contain less contaminants than their dry counterparts.

There are three limitations of liquid fertilizers. The first limitation is the fact that they are produced in relatively few grades, usually 9-18-9, 10-20-10, and 3-18-18. If soil testing indicates you need another grade, you simply can't provide it with the liquids available. A second limitation is that they require liquid equipment to handle. If your equipment is set for only dry products, you will need to spend time and money converting to liquid application. Given the advantages of foliar spraying with liquid fertilizers, however, it will be well worth your time and expense.

The third limitation has to do with the constant use of the same product on the same field. As Dr. Reams taught, energy is released by the interaction of one nutrient against another, specifically calcium. By using the same fertilizer grade year after year, the field gradually begins to vibrate with the same energy beat as the fertilizer. This means more and more fertilizer is needed to produce the same result. Different energy sources are needed to break out of this cycle. Applying lime and changing fertilizer grades may solve the problem.

Dry Soluble Fertilizers

The newest type of fertilizer being marketed today is the dry soluble, a greenhouse grade material being sold for general farming use. These products have the advantages of both dry and liquid products. The products are shipped dry so the farmer doesn't pay for the transportation of water. He then mixes them into his own water on his own farm for application. Because dry soluble fertilizers are usually made from refined raw materials, they generally contain high levels of energy. For this reason, farmers have been successful using them in amounts ranging from 5 to 10 pounds per acre preplant, 5 to 10 pounds sidedress, and 1 to 5 pounds foliar sprayed.

Many companies marketing dry soluble fertilizers offer as many as 10 fertilizer grades which can usually be mixed together. This enables the farmer to obtain good accuracy in terms of applying the nutrient combination recommended by soil testing and needed by that particular soil or crop. Lastly, dry soluble fertilizers are usually made from highly refined, technical, or food grade raw materials which assures cleanliness.

Dry solubles can have limitations in hard water areas. The laws of chemistry only allow so much dissolved material in a given amount of water. Lowering the water pH with vinegar, raising it with baking soda or household ammonia, or heating the water may help solubility problems. Dry solubles which contain calcium can also be problematic if other materials added to the solution contain soluble phosphates or sulfates. When mixed with calcium, they form insoluble salts of calcium phosphate (hard rock phosphate) and calcium sulfate (gypsum).

Dry soluble fertilizers can also have limitations in field response. Using a low-input, "high-energy" program requires other soil factors to be ready for such a situation. Soils with a strong imbalance in the calcium/magnesium levels, soils which have very limited microbial activity, and soils whose owners/managers are unwilling to apply the prescribed formulations at the times such are prescribed to be applied are all candidates for not using dry soluble fertilizers.

Contamination

All fertilizers, whether dry, liquid, or dry soluble, can contain contaminants which may be detrimental to the soil, crop, groundwater, or farmer. These can be picked up through the mining operation, the manufacturing operation, or from poor handling procedures. Recent revelations have shown a deliberate use of radioactive material as a filler in dry fertilizers in an attempt to use farm soil as a dumping ground for this material as well as to avoid the cost of proper disposal. Prior to the electronic scanner the farmer was at a loss to identify either product contamination or the effect of the contamination on plant or soil. Now he has the capacity for both.

Using the electronic scanner, the farmer can measure not only the contamination in the products he intends to purchase, but he can also measure the effect these products will have on his soil or crops prior to purchasing them. By obtaining product samples from the manufacturer, the farmer can analyze his soil sample, compare with the fertilizer samples, and make management decisions accordingly.

Reducing contamination allows the natural soil processes to occur and increases the chances of plant/crop response.

Chapter 11
Selecting the Correct Fertilizer

Background Considerations

Chemical fertilizers are generally considered to be water soluble and can be detected after the application by measuring solutions of soil water. Since they are water soluble, they are subject to leaching, dilution, and loss of movement during a drought. Ideally, bacteria will consume these fertilizers and hold them in the soil in the form of their own protoplasmic remains. Once this happens, the fertilizer is now considered stabilized and is not subject to leaching. It is available in a biologically active soil for the needs of the plant. It is extremely important, therefore, to use fertilizers and other chemicals which will *feed the soil* and not just feed the crop. This has been one of the main issues behind the organic farming movement.

Although chemical fertilizers alone can be used to feed plants (hydroponic farming consists of growing plants in a water solution with no soil support), using them in this manner results in increased farming costs. A much better idea is to feed both soil and crop and receive cooperative nutrient support from the soil as well as from the fertilizer. The primary task for the farmer is to correctly select the right fertilizer for his fertility needs.

Fertilizer purchases make up a major line item in any farming operation and improperly selecting fertilizers or applying too much or too little can cause soil imbalances, but can also be financially draining. What basic fertilizers are available today? What are their merits? Which are the best to use?

Fertilizer selection is best done with some scientific information as a guide. The farmer is advised to have proper soil testing performed on his fields prior to selecting fertilizer. Often he finds that his local elevator doesn't carry particular fertilizer numbers or products. The pressure always exists to sell profit-making items, those generally recommended or purchased in the local area. However, the farmer must also realize that it is he who is making the purchases, and he needs to be assertive as to what he is or is not going to purchase.

Dolomitic limestone may be commonly recommended, but if enough farmers refuse to purchase it, the lime dealer will begin to offer high-calcium lime. As a purchaser the farmer not only has the right, but also the responsibility, to purchase only those products that he feels will be best for his needs. He should demand to see certification or guarantees of ingredients (these are required by state laws) before the purchase. Buy what you want, not just what the elevator has to sell.

Nitrogen

Nitrogen makes up 80 percent of the air we breathe. This should be the major source of nitrogen for agriculture obtained through several processes. Electrical storms convert nitrogen to nitrogen oxide or nitric acid which is then carried by rain to the soil. Rhizobia bacteria, which form on the roots of legumes, are capable of fixing nitrogen from the air as are some forms of algae, especially the blue-green varieties. In recent years, field examination of legume roots have found little or no nodulation which may indicate that the Rhizobia cannot survive in the "growth medium" found in those soils.

Soil organic matter often contains 5-6 percent nitrogen, much in the form of protein. This needs to be broken down by soil microbes to become available to the plant. Ammonias released during this process are held in the humus. Soils low in humus do not hold nitrogen well.

Nitrogen is considered a primary plant nutrient. It is a constituent of every living cell and is associated with the production of vegetative growth and a dark green leaf color. Lack of nitrogen, generally recognized through the light green coloration in plants, is thought to be associated with a lack of chlorophyll in the leaves. Nitrogen is needed in both the nitrate and the ammonium forms. Nitrogen is an essential component of amino acids, proteins, and other forms of basic cell building blocks.

Soil nitrogen is naturally associated with the organic component of the soil. The higher the organic matter or humus content, the higher the amount of not only nitrogen, but of all other beneficial soil and crop nutrients held. In fact, soil microbes consume fertilizers as well as organic matter and hold them in their bodies (primarily protoplasmic fluid) as a highly insoluble, but available, form of plant food. High organic matter content in the soil makes for a more efficient utilization of any additional nitrogen added by the farmer.

The breakdown of ammoniacal nitrogen (NH_4) into nitrate nitrogen (NO_3) is a major cause of soil acidification and loss of nitrates through leeching. Soil pH is lowered and lime is required to raise it. Applying excessive nitrogen, therefore, pressures soil pH downward. Additionally, the application of large amounts of nitrogen, 40 pounds or over, usually overrides the microbial production of nitrogen. By adding large volumes of nitrogen at one time, the farmer is actually shutting down the natural microbial response which will produce free nitrogen.

Considered from an energy perspective, nitrogen acts like an electrolyte. A farmer who owns a tractor with an old battery can keep putting more acid (electrolyte) into the battery to get that last cranking charge before buying a new one. By adding sulfuric acid, more electrolyte is added and one last crank may be obtained. Nitrogen serves a similar role in the soil. Nitrogen acts as an electrolyte allowing energy (nutrients) to "flow" through the soil. By adding more nitrogen, the farmer is forcing the soil to give up more energy in the form of nutrients from the soil colloid or soil solution to the plant. This approach works both in the tractor and in the soil, but only for a limited time before the system begins to break down. A new battery can easily be purchased, but soil repair will be needed unless you can afford to buy new land.

Sources of Nitrogen

Anhydrous ammonia NH₃ (82-0-0) — This is a highly volatile gas contained in sealed cylinders and is injected into the soil with special applicators. Farmers may also bubble it through water and apply it as a liquid nitrogen solution called aqua ammonia. Adding a carbohydrate or sugar source to the aqua ammonia will greatly increase its stability and efficiency.

Anhydrous (meaning without water) ammonia is very dangerous and must be handled with extreme care as it can burn the skin through dehydration. The gas readily draws moisture into its chemical structure when it comes into contact with the skin or eyes causing a burn. This product is extremely hard on humus and organic matter found in the soil as well as on beneficial soil microbes. Stories are told of how it was used to create airplane runways in jungle soils during World War II. It was injected into the soil in such high concentrations that it would burn out all organic matter leaving the soil hard and compact.

Anhydrous is easily lost into the atmosphere as witnessed by white puffs of vapor following an applicator in operation. Since it contains the ammonium form of nitrogen, it is actually the wrong form of nitrogen to be applied in the early spring. Although it will produce a crop, it should generally be eliminated from biologically oriented programs. If you are intent on including anhydrous in your programming, you would do well to bubble it into water to make aqua ammonia water and include a form of carbon, such as sugar or molasses, to help "hold" the ammonia once it's applied. Anhydrous generally lowers soil and plant energy levels. It is one of the main products to avoid.

Urea (46-0-0) — Urea is a dry, synthetic form of nitrogen similar to that found in animal urine. It is one of the least expensive forms of nitrogen and its use is widespread. It is also used in livestock feeds as a non-protein nitrogen source supposedly to raise protein levels. Urea may tend to dry soils out and overuse can lead to soil imbalances (as can any product). It can be effective if used in moderation. It should always be worked in or watered in as it is very volatile. Urea shows up well in foliar applications because the plant can use the urea nitrogen with little or no internal energy use. The energy level of fertilizer grade urea is not very strong which the authors believe is related to formaldehyde residue from the manufacturing process.

Liquid nitrogen (28, 30 and 32%) — These liquids are usually a mixture of urea and ammonium nitrate in a water solution. They are generally good products to use, supply the soil with both nitrate and ammonium forms of nitrogen, are less harsh on plants, and basically are a good buy for the money. Do not confuse these liquids with a cheap ammonia water (aqua ammonia). The odor should reveal the difference. Quality products have a slightly oily, viscous look. Energy readings can be good.

Aqua ammonia (21-0-0) — Aqua ammonia is ammonia water. This form of ammonia evaporates less than anhydrous ammonia, but is still not very stable. It is best to add some sugar or molasses to the solution to aid in nitrogen retention. Partial amino acid compounds may be formed in solution, and increased bacterial stimulation will enhance uptake and conversion to a non-leachable form in the soil. Depending upon your carbohydrate source, this product will vary in its energy readings.

Ammonium nitrate, NH_4NO_3 (34-0-0) — This is a dry, granulated product containing both ammonium and nitrate forms of nitrogen. Ammonium nitrate releases its nitrate content first and its ammonium content later. Ammonium nitrate can be explosive, but in most farming operations, the product is safe to store and use. It is used to make liquid nitrogen solutions and often has good energy readings when applied in its dry or liquid forms.

Ammonium sulfate $(NH_4)_2SO_4$ (21-0-0-24S) — A dry product, ammonium sulfate can be obtained in a variety of colors.

The darker, grayer colored form, a by-product of the nylon industry, works best as a fertilizer. This product is good to use both for its nitrogen as well as sulphur content. Some alternative agriculture authorities feel the product can work as a temperature control regulator for soils with active humus and calcium. This soil becomes less prone to high temperatures in the summer and to low temperatures in the late fall and early spring. It can be an excellent product when used at lay-by time and, depending on the source, can have good energy levels.

Ammoniated phosphates — Ammoniated phosphates comprise two dry forms of products commonly known as DAP (diammonium phosphate), $(NH_4)_2$, HPO_4 (18-46-0), and MAP (monoammonium phosphate), NH_4 H_2PO_4 (11-52-0). Their advantage lies in the combination of nitrogen with phosphate as will be explained later. Although DAP has more nitrogen than MAP, it is observed that DAP sometimes lowers soil energy levels while MAP often raises them.

Calcium nitrate, $Ca(NO_3)_2$ (15-0-0-19Ca) — This is an excellent dry source of both nitrate nitrogen and calcium. Although it is considered a fairly expensive form of nitrogen (it is imported from Norway), it has been shown that calcium nitrate will aggregate (fluff up) the clay colloids in a desirable way and increase soil pH as well. Calcium nitrate has a good energy reading.

Potassium nitrate, KNO_3 (13-0-46) — This dry product would be used more for its potash content than its nitrogen, although both are good. It is used heavily in the vegetable, potato and tobacco markets. Also called *saltpeter*, it was used in the manufacturer of gunpowder. It can have excellent energy levels.

Manures — Manures have long been acclaimed by organic people as essential to good fertility practices. Government opinion has ranged from "they are not needed" to "use them if you have them available, but you can get along fine without them." Manures are now recognized as an excellent way to increase the soil cation exchange capacity ratings and to make the addition of added fertilizers more efficiently utilized by the soil.

According to the University of Pennsylvania, the total amount of nitrogen which is readily available the first year plus that which is available in subsequent years from manure is as follows:

Animal	Waste Type	% Nitrogen
Beef	without bedding	0.55
	with bedding	1.05
Dairy	without bedding	0.45
	with bedding	1.45
Poultry	without litter	1.65
	with litter	2.80
Swine	without bedding	0.50
	with bedding	1.40

Manures add primary nutrients, but their main uses are the addition of organic matter, their bacteria feeding qualities, and the trace nutrients they contain. Manure is almost 100 percent available plant food when properly decomposed. Energy readings will vary depending upon the manure source, contaminant level (such as antibiotics), and stage of decomposition. One major problem with manure is that it is often putrefied due to improper handling in most conventional operations. Manure which smells bad, attracts flies, and doesn't decompose is bad for the soil. Better quality feed going into the livestock and better handling will help.

Legumes — Legumes are crops which will symbiotically attract a Rhizobium (bacteria) species to produce root nodules which biologically fix nitrogen from the air into the root nodule. Edwin McLeod has an excellent book, *Feed The Soil*, which goes into detail describing how the legumes feed the soil and then, subsequently, feed the plant. He discusses legumes appropriate for different soil types and conditions. He lists seven major groups of beneficial legumes: alfalfa, bean, cover, cowpea, lupine, soybean, and vetch.

Although many legumes are primary forage or feed staples, they also make excellent cover crops. These cover crops, also known as "green manure" crops, are important in many ways other than just their nitrogen-fixing abilities. They play a crucial role in soil aeration, erosion control, and crop rotation, to name a few.

Green manure crops are an important consideration for cash crop farmers who do not spread animal manure. Green manure crops can be used to stimulate soil bacterial levels, to make nutrients more available to plants, and to reduce erosion through increased root mass. Prior to plowing down a green manure crop, it may be beneficial to spray the crop with a bacterial product, a carbohydrate, and a nitrogen source to assist in decomposing the crop.

Tankage — Tankage usually consists of rendered, dried, and ground waste of slaughtered animals, i.e., hides and blood. This form of nitrogen is, like the manures, used mostly by organically inclined farmers. Its major importance may be its effect on soil microbes. It can contain contaminants of heavy metals and antibiotics. Energy levels will vary according to source.

Liquid fish — Liquid fish is a by-product of the fishing industry and is similar to tankage in its beneficial effect on soil microbes. Fish contains low levels of nitrogen (about 4 units or pounds per 100 pounds), however, it is an excellent protein source to use for building soil microbial levels. Coupled with soil bacteria, it greatly assists in the formation of amino acids and protein development. This product can have very good energy levels.

Phosphorus

Phosphorus is considered a primary nutrient which promotes plant growth, hastens maturity, and stimulates seed development. It's contained in every living cell and is considered critical to the process of photosynthesis and energy transfer. Phosphorus is often plentiful in commercially farmed soils, but in forms which are tied-up and not available to plants. Lower or higher soil pH's will tie up phosphorus forming tricalcium phosphate at higher pH's and aluminum or iron compounds at low pH's.

Dr. Reams felt that phosphorus was crucial to the plant's growth mechanism. He taught that the mineral needs of plants require that all minerals except nitrogen must be present in the phosphate form for optimal growth response. He stated, "Upon passing from the root zone to enter the plant stalk, a crucial line exists called the pons." Here the plant electrochemically fixes the nutrient in question with phosphate so it will go to a particular part of the plant, for instance, leaf, stalk, or fruit, based on the new frequency established at that point. Through the growth process, the phosphate would then travel back down through the stalk, re-enter the roots, and the cycle would repeat. This cycle gives a clue as to why phosphate is not necessarily required in large quantities (it cycles), yet it is important to be available in sufficient amounts.

In Chapter 14 of this handbook, we have reprinted a chart of refractive indicies of crop juices which lists refractometer readings for many fruits, grasses, and vegetables. According to Dr. Reams, phosphate is one of the key nutrients required to produce these higher sugar readings. Contrary to most soil lab recommendations which give a 1:2 phosphate:potash ratio, Dr. Reams wanted the LaMotte phosphate:potash ratio to be 2:1 on most soils and 4:1 on fields growing forages and grass-related crops. These levels of phosphate will provide the energy production to allow the plant to develop the sweetness and specific gravity inherent in its genetic makeup. This ratio is also important for broadleaf weed control as mentioned in Chapter 4.

Most farmers immediately recognize, however, that the dry standard phosphate fertilizers sold today could never be used to build these types of ratios. The phosphate in 0-46-0 would tie up with calcium and other soil minerals within 30 to 60 days, or sooner, and would be unavailable to the plant. For this reason, Dr. Reams made extensive use of colloidal, or soft rock, phosphate. The colloidal phosphate does not tie up due to its colloidal nature, and it contains trace amounts of many other minerals in colloidal form.

A conventional approach to fertilizing with phosphate is to apply amounts according to standard soil test results. In the case of phosphate, the problem with this approach, as discussed in Chapter 9, is that 0-46-0 is not necessarily available to the plant. If measured by an electronic scanner, you would find the energy level to be low, indicating less than optimal performance. It is at this particular point that Dr. Reams differed so radically from conventional approaches.

The "secret" to making plant nutrients available for use by the plant such that the plant can grow optimally is for the nutrient to be in a *biologically* active, not *chemically* active, form. Once microbial life consumes the nutrient, fixing it in humus, the material becomes biologically active. This distinction means the plant can now utilize the materials much easier than otherwise, the soil will begin to balance itself naturally because of the microbial action and such variables as increased specific gravity, resistance to stress, and even weed control will be forthcoming.

Phosphate does not move easily in the soil. It is not a mineral which needs to be applied on a constant basis once the base is established. However, until the soil becomes biologically active, it is usually necessary to add phosphorus. Often only small amounts, such as 3 to 5 units per year of highly available forms will suffice is used in the row as a starter. It can be applied in every foliar application (be cautious with calcium) in order to assure minerals enter the plant in the phosphate form.

Sources of Phosphate

Triple superphosphate (0-46-0) — This ingredient is used in making many dry fertilizers. It consists of hard rock phosphate treated with phosphoric acid. It is a fairly "hot" form of fertilizer which supplies lots of energy — too much, in fact. Its energy release is initially too strong for young plants and its chemical activity is such that it will usually tie up with calcium within 10 to 60 days after application making both nutrients unavailable to the plant. It ties up most readily in the higher pH ranges and often becomes toxic to micorrhyzia (soil fungi).

The authors are now finding soils with extremely high CEC phosphate levels, good LaMotte phosphate levels and low plant uptake. This contradicts the usual Reams theory and may be attributed to the total absence of micorrhyzia to move the available phosphorus into the plant. It is not uncommon to utilize only 10-15 percent of 0-46-0 in two years' time which makes it relatively expensive to use even though the initial cost may be moderate. It is because of this phosphate tie-up that soil testing labs recommend relatively large amounts of phosphate even though they also may report relatively high levels in the soil. They are covering their tie-up bases. This product has a high energy level; however, it usually doesn't test well against fields or crops. This is one of the main products to avoid.

Superphosphate (0-20-0) — This dry product preceded 0-46-0. It is a rock phosphate treated with sulfuric acid which makes the phosphate more readily available plus provides a source of sulfur as gypsum. This product is practically unavailable today. It is a better source of phosphate than 0-46-0, but is a lower grade and isn't as profitable to handle. It also isn't as reactive, which means it doesn't tie-up as readily. It has a good energy reading and would be good to use judiciously if obtainable.

Polyphosphate (10-34-0) — Generally a good product to use, this liquid is utilized for immediate phosphate needs as a starter. Energy levels are often good, but check for contamination — heavy metal and otherwise — from use of industrial acids used as raw materials.

Ammoniated phosphates (DAP 18-46-0 and MAP 11-52-0) — These dry materials are made by combining ammonia with superphosphate. Most electronic scanner operators recommend the use of MAP as a cost-effective alternative to supplement the use of colloidal phosphate. MAP works well in the soil, tends to be less costly, and may help raise phosphate levels. This can be an excellent product to use for short-term benefits. DAP is generally found not to benefit the energy levels of soil, microbes, plants, etc., but may be useful in some situations. MAP usually has good energy readings.

Phosphoric acid (85%) — Phosphoric acid is a highly refined, liquid form of phosphorus which can be very beneficial. This product is commonly used in making liquid fertilizers. It can also be used in self-made fertilizers, however great care must be taken in handling as it is a highly oxidative material which can do great damage to metals, clothing, eyes, and skin. *Always* add phosphoric or any other acid to a tank of water and *never* add water to straight acid. Phosphoric acid has a good energy reading unless it has been spent or used prior to wash down metal. Spent acid usually contains heavy metal contaminates.

Hard rock phosphate (0-3-0) — This is a mined (32%) rock phosphate which contains a broad spectrum of trace minerals as well as calcium. Most deposits are in a relatively insoluble form and will last for many years while being broken down by microbial life. Its availability increases with increases in beneficial soil microbial action. Most hard rock phosphates are found in Idaho and Montana, with those in Montana having somewhat less of the very heavy metals such as cadmium. Energy levels can be good if the ratios of all minerals are taken into account. It can be considered for use in low phosphate level soils.

Colloidal phosphate (0-2-0) — This product, commonly called *soft rock phosphate*, is a by-product of the hard rock phosphate mining industry. The colloidal clay which is washed off the hard rock material is collected in large ponds and later sold as soft rock phosphate. This product is much more available to plants and soil organisms than its hard rock parent and provides scores of highly beneficial trace minerals. It is relatively slow releasing and will not present the tie-up problems characteristic with hard rock phosphate treated with an acid. It contains a relatively large amount of calcium along with an 18-22 percent phosphate content.

Reactive rock phosphate (0-3-0) — This is another byproduct of the hard rock industry. It is washed off with water and left in above-ground piles. The only deposit currently available in the United States is called "Tenn. Brown Phos." It contains less total phosphate than hard rock, but more than colloidal. (22-25%). It contains a very broad spectrum of trace minerals, as well as about 20% calcium. Because it is geologically classified as a reactive rock, the actual availability may range from 5 to 12%. It also contains much less of some of the not-so-desirable traces such as mercury and cadmium.

Low price, cleanliness, location and organic status have made it one of the best-selling phosphates in the United States. It usually spreads easily with a spinner spreader.

Bone meal — This by-product of the animal industry supplies phosphate from animal bones. Bone meal is generally used by organic proponents. It can contain heavy metal contamination, however, including airborne lead ingested by the animal source. Energy readings will vary and cost per unit may be prohibitive for larger acreage.

Potash

Potash is needed to build strong stalks of sizable diameter, it controls the caliber (size) of the fruit, increases the yield of tubers and seed, and is necessary for the plant to produce starch, sugar and oils. It can be obtained from the air by many plants when the total soil ecosphere is working properly.

Potassium is present in soils in the form of silicates which are insoluble. Many soils have enormous potassium reserves, and biological activity can make it available to the plant. Soluble potassium is held onto the negatively charged clay colloids. As conventional practice has emphasized the use of potash it has often overloaded the clay colloids, thereby taking the place of calcium.

Potassium does not have the same electro-chemical properties as calcium and does not provide the same support to the clay structure. The excessive potassium can result in structural collapse of the soil which can affect the fertility and increase compaction. Potassium also will increase the soil's pH, which is one of the reasons not to use pH as an indicator in determining the need for liming.

Sandy soils can be particularly difficult to work with in maintaining potassium levels. Having little or no clay content, they do not hold potassium well. Potassium will easily leach out of the root zone and be lost.

The most prevalent form of potash used and recommended today is muriate of potash (0-0-60). It is recommended by universities for heavy feeding of alfalfa and may be the single most detrimental product used on the farm today. Its use will cause growth to occur, but the alfalfa will produce hollow (empty) stems, will bloom before potential growth has occurred, and will be high in both nitrates and potash. This amounts to "gunpowder" (potassium nitrate) hay.

The hay might also be low in calcium and phosphorus requiring the farmer to "go downtown" to buy expensive calcium and phosphate feed additives. Even if the cash crop farmer sells his hay and doesn't feed it, the loss of revenue due to light hay with hollow stems should cause alarm and lead to other fertility programs promising a higher quality result. As one Amish farmer said, "I get as much feed value out of one bale of this man's hay compared to three bales of that man's hay. I prefer the former and pay a premium for it."

Dr. Harold Willis in *The Rest Of The Story* goes into detail about soil life, fertilizer nutrients and their effects on soil life. He has documented the effects of muriate of potash on the soil. According to *Fertilizer Technology & Use*, the culprit in muriate of potash is the chloride ion which comprises roughly 40-47 percent of the product. Whereas only 2 parts per million of chlorine is enough to sterilize drinking water, 50 to 200 parts per million of chloride are added to farm soil when applying 200 to 800 pounds per acre. One hundred pounds is equivalent to applying 100 gallons of Clorox bleach to the soil.

Killing bacteria in drinking water may be appropriate, but what possible rationale could be given for the destruction of beneficial soil microbial populations? Potassium chloride (muriate of potash) may be cheap in the short run, but it is very expensive over the long haul.

Sources of Potash

Potassium chloride, KCl (0-0-60) — This is the most commonly recommended and used source of dry potash in Canada and the United States. It is considered the cheapest source. Potassium chloride comes in several grades including 0-0-59, 0-0-60, and 0-0-62. Although proponents claim the chloride content varies with the grade, the fact remains that potassium chloride comprises molecules containing chloride and the percentage difference between them is negligible. This product seldom tests good for soil application and is one of the main products to avoid.

Potassium sulfate, K_2SO_4 (0-0-50) — Also called *sulfate of potash*, this dry product has very little chloride and is an excellent substitute for muriate of potash. It often costs as much as three times that of potassium chloride. However, at three times the price, it is still a bargain because it contains five times the calculated energy. Potassium sulfate usually has a good to excellent energy reading.

Natural Potassium Sulfate Plus (0-0-7-15S-1Fe) — This is a natural potassium and sulfur complex from an ancient volcanic plume. It is noted for releasing tied-up K in the soil. It contains multi-oxidation states of S and active sulfating bacteria. Therefore, it is very fast acting and effective in helping high pH or high magnesium soils. NKS-Plus contains lots of trace minerals and has very high energy readings.

Potassium nitrate, KNO_3 (13-0-46) — This is a good, highly energized form of dry potassium. It is commonly used in growing vegetables. The best material is imported. It has an excellent energy reading.

Chilean nitrate of soda potash (15-0-14) — This dry product has a good to excellent reading although it can have a high sodium content. It is a mixture of sodium and potassium nitrate.

Potassium magnesium sulfate (Sul-Po-Mag 0-0-22-18Mg-22S) — This is an excellent natural dry source of potassium, magnesium and sulfate in a complex mineral form called langbeinite. Dr. Reams claimed it works best in the northern hemisphere when applied between July 15 and September 15. During this time, it supposedly works to release copper which allows plant bark to expand and stretch. This is a great product for use on orchards, and it will also work well on farm crops. It has a very good energy reading.

Greensand (7%) — Greensand (glauconite) is a natural, mined, dry potash source for organically inclined farmers. It is relatively insoluble in water which means it won't leach and will require good bio-activity to be made biologically available. The energy reading is fair to good.

Manures, tobacco stems and wood ashes — These can be good sources of organic potash, but they can be contaminated with toxic sprays, antibiotics (used in animal care), and lead from paint. Be careful not to overuse, thereby increasing weed pressures. Energy readings will vary according to the source.

Kiln dust — This by-product of the cement industry can have significant amounts of potassium, calcium, sulfur and trace minerals. Energy content is usually excellent.

Calcium

Calcium is basic to every living cell. It contributes to cell wall strength and helps to regulate many cell processes. Additionally, it is very important to soil micro-organisms and soil structure. Soil lacking in calcium tends to produce aldehydes as a result of improper decomposition due to poor aeration and loss of structure. The aldehydes are what can embalm or preserve organic matter instead of turning it into humus. Plowing up corn stalks plowed down two years previous is a good example of this embalming effect.

Calcium has, until recent years, been considered a secondary nutrient. In 1981, the authors of *Fertilizers and Soil Amendments* stated, "Calcium, and sulfur are designated secondary elements because . . . they are generally not added as fertilizers in large quantities. The term *secondary* may not be appropriate because there is an increasing tendency [for them] to be grouped...as major nutrients." These authors, from Kansas State and Michigan State universities, along with others, are now rewriting the textbooks regarding our understanding and realization of the importance of calcium in soils.

New findings in other fields are verifying calcium's importance in plant and animal cells for support and functioning. According to Dr. Reams, calcium is the single most important mineral needed for plant, soil and animal nutrition by weight and by volume. Soil pH has been used as the primary indicator of lime needs. Since other positive ions, such as potassium, magnesium and sodium can also raise pH, lime applications have not been made in many situations that really were in need of the calcium component.

The standard source of calcium for soil for centuries has been calcium carbonate. In the authors' experiences, application of high-calcium lime to a soil above 7.0 pH has sometimes actually lowered the pH due to the complex biological and chemical processes found in living soil. A non-toxic program calls for viewing soils as to their available calcium content, rather than using the pH concept.

Besides the actual amount of calcium, the ratio of available calcium to available magnesium is also critical. If the range is too narrow — less than 7:1 — there is a tendency for the soil not to hold nitrogen. As ratios of 2:1 or 3:1 are approached, it is very difficult to keep enough nitrogen available to grow a crop without more frequent applications or larger amounts of actual N. Low-level applications of N and biological fixation would probably not be sufficient to provide a high nitrogen user like corn to react satisfactorily.

As with all fertilizer materials, Dr. Reams approached calcium from an energy concept. Using the energy concept, he stated that 80 percent of the energy necessary to grow a crop comes from nutrients carried along the magnetic path from the south to the north pole in the air. These nutrients include minerals, carbon dioxide, water vapor, sunlight, cosmic radiations, plus trace nutrients and potash. The remaining 20 percent of the energies comes from the interaction of soil materials with calcium. And 80 percent of this 20 percent (16 percent total) comes from the energies of calcium. The remaining 4 percent comes from the energies associated with other nutrients, such as N, P, K, S, and traces found in the soil. In the current plant physiology texts, 90 percent of the elements that end up in the plant are credited as coming from the air and 10 percent from the soil.

It is because of this understanding of energies that Dr. Reams stressed liming and the measuring of calcium levels through the CEC and LaMotte soil tests. Dr. Reams built his fertility programs on calcium energies complemented with phosphate energies.

Calcium energy forms the basis from which other nutrients (energies) benefit plants. Sulfur products, for instance, will work much better (supply more energy) if adequate levels of calcium are present. In fact, sulfur in sulfate form is often used to *kick* calcium energies loose. Some of the effects observed when using anhydrous ammonia are related to the release of calcium. Calcium can be present, but tied-up, with phosphorus or other elements. These tie-ups reduce the energy available for plant growth. Using soil tests, an experienced soil or crop consultant can determine whether or not calcium (or any nutrient) needs to be added or released.

Different lime deposits contain different energy levels. Using the electronic scanner the consultant can find that one source of lime may have much more energy available than another source. A quick comparison of dolomite to high-calcium lime will usually show the high-calcium lime raises the general vitality of the plant much higher than does dolomite. The most cost-effective approach is not necessarily purchasing the least expensive product, but rather the product which best raises the energy levels of a field and crop.

Sources of Calcium

High-calcium lime, $CaCO_3$ (25-35% Ca) — This dry product, also called calcitic limestone or calcium carbonate, is usually the best, most economical source of calcium. It is a mined material which is generally available in agricultural areas. Usually recommended for raising soil pH, it is fairly inexpensive to purchase and have spread. High-cal lime generally contains less than 5 percent magnesium. This is the preferred lime in most situations. Energy levels will vary.

Dolomitic lime, CaCO₃ MgCO₃ (6% Mg minimum) — Also called dolomite, ag lime or magnesium limestone, it contains a relatively high amount of calcium as well as significant amounts of magnesium. Dolomite is recommended to correct calcium and magnesium deficiencies. Depending upon magnesium levels, dolomitic lime can often cause more harm and trouble, however, than its initial cost. Excess magnesium is associated with soil stickiness, crusting, compaction, reduced aeration, and releasing nitrogen from the soil pound for pound. It can also cause both phosphorus and potassium deficiencies, lower the availability of calcium, allow organic matter to form aldehydes which kill beneficial soil bacteria and take the place of calcium in plants and soils which causes poor quality crops. Energy levels will usually be lower than high-calcium lime when compared with the soil or plant. This usually is a form of lime to avoid.

Marl — Found locally, this material is generally purchased for a low price and contains varied levels of clay, peat and shells. It can be a good, inexpensive source of calcium and other minerals if energy levels are good.

Hydrated lime, Ca(OH)₂ — This consists chiefly of calcium and magnesium hydroxides. It can be a good product for emergency energy needs. It is a strong alkali and must be used with caution.

Quick lime, CaO (46% Ca) — Also called calcium oxide, this dry product is very fast acting, contains readily available calcium and is loaded with energy. Use with caution or you can burn crops.

Gypsum, CaSO₄ — This dry form of lime may actually lower the soil pH because it contains sulfate sulfur which gives its souring or acidic qualities. It loosens the soil and is a good source of lime to use on high pH and sodic soils. It can have good energy readings.

Liquid calcium (8-0-0-10Ca) — A relatively new product on the market, liquid calciums usually contain glucoheptanates (corn sugars) or other chelating agents in conjunction with calcium and nitrogen. This type of product has proven to be useful in making calcium rapidly available. Energy levels will vary according to manufacturer. (Liquid lime is a water slurry containing insoluble calcium carbonate or other forms of lime and is not the same as liquid calcium.)

Magnesium

Magnesium, like calcium, is now being considered as a primary nutrient. It is an integral part of chlorophyll making it essential for photosynthesis. Enzyme systems involved in energy transfer and respiration do not function properly with low magnesium levels. Magnesium can be related to yield. Magnesium is a difficult nutrient to assess because of the many negative effects it has when applied to the soil in excess (see Dolomite above). Magnesium can be picked up by the plant from the air when the total ecosystem is in good harmony.

Sources of Magnesium

Dolomitic lime — see Calcium above.

Magnesium sulfate, $MgSO_4$ — Also referred to as epsom salts. It is an excellent source of magnesium when soil or plant tissue analysis shows magnesium to be needed. This product usually has good energy readings, is highly available, and does not carry the negative aspects of dolomitic lime. It is used as a foliar spray or for ground application and can help crops release excess nitrogen.

Magnesium chelates — Liquid chelated magnesium products often have excellent energy readings.

Sulfur

Sulfur, like calcium and magnesium, is now considered a primary nutrient. Sulfur is used by plants in the sulfate form. With the discontinuation of using 0-20-0 in favor of 0-46-0, the farmer found that a common source of sulfate sulfur was no longer available. Sulfur is needed in protein and amino acid formation, in the formation of nodules on legumes, and in many other plant processes. It is also used, both in combination with a calcium product or by itself, to make calcium energies available to plants. Sulfur plays an important role in any fertility program and is often used as a sidedressing at lay-by time as part of a sulfate compound such as potassium thiosulfate. Most soil tests report sulfur as being high at lower levels than we recommend. Good liming and high sulfate levels make for a dynamic, living, growing system.

Sources of Sulfur

Ammonium sulfate (21-0-0-24S) — see Nitrogen above.

Sul-Po-Mag (0-0-22K-18Mg-22S) — see Potassium above.

Gypsum (calcium sulfate), CaSO₄ — see Calcium above.

Ammonium thiosulfate (12-0-0-26S) — This is an excellent source of high energy liquid sulfate-sulfur in a modified ammonium form which works well on the soil and also as a foliar spray. It can have good to excellent energy readings.

Natural Postassium Sulfate Plus (0-0-7-15S-1Fe) — see Potassium above.

Trace Nutrients

Trace nutrients have only recently come into their own. For years, farmers were unaware of the need for, or importance of, trace minerals in the plant or soil. Consequently, they literally mined their soil of valuable trace minerals. Organic adherents covered trace mineral needs through the addition of seaweed (kelp) and fish products, and rock minerals such as soft rock phosphate, glauconite and granite dust, which contain most of the required trace minerals although in very small quantities.

Because trace minerals are crucial to a healthy plant and soil their lack has resulted in a dramatic decline in plant and soil health and corresponding food value. Witness the low sugar (brix) refractometer readings which characterize most agricultural crops today. This lack has been the reason why farmers have been forced to purchase an increasing amount of costly mineral supplements for their animals. It is, in our opinion, also the reason why foreign countries are refusing to purchase low-quality American crops and why so many experts recommend nutritional supplements to be taken with meals. Trace nutrients are called "trace" because they are required in small quantities. Also called "micronutrients," they officially include: boron (B), calcium (Ca), chlorine (Cl), cobalt (Co), copper (Cu), iron (Fe), manganese (Mn), magnesium (Mg), molybdenum (Mo), nickel (Ni), sodium (Na), sulfur (S), and zinc (Zn).

These have been identified as being important to cell or plant growth and development by the Association of American Plant Food Control Officials, the fertilizer regulatory officials. How many more minerals, such as selenium, fluorine, arsenic, silver and iodine, are important or crucial remains to be seen as technology provides the capacity to identify new minerals and the roles they play.

Trace minerals are the key to plant health and insect resistance. Consultants promoting the use of a refractometer to measure plant sugar content explain that in addition to calcium and phosphorus, trace minerals, in proper ratios, are required to raise the sugar levels of crops. This raised value is also associated with higher protein content, higher mineral content, increased sweetness, and a higher specific gravity. In addition, these crops will have a lower nitrate and water level, will store better, and will withstand lower air temperatures.

Many biologically oriented companies market fertilizers containing fish and seaweed. Seaweed contains all the naturally occurring trace nutrients known to man and has been used for centuries as a soil amendment by farmers living near the ocean. Fish has also been used for centuries by farmers (including American Indians) for fertility purposes. These are excellent sources of trace nutrients. Although they are not well adapted to supplying primary nutrient needs, they offer an excellent way to stabilize and extend the availability of chemical fertilizers or replace them as starters.

Trace mineral availability depends upon two main variables. Soil pH can have an influence on availability. Microbial activity is the second variable. When microbes are functioning optimally, the trace minerals will usually be available.

Sources of Trace Nutrients

Trace nutrients come premixed in fertilizers, can be requested as additions to custom mixes, and can be purchased in both dry and liquid forms. Most can be obtained in the sulfate form, as found in copper sulfate or iron sulfate, or in the oxide form as found in magnesium oxide. These are the most popular and least expensive forms. These forms, however, aren't of the highest energy nor are these the most biologically available forms. Other forms such as amino, citrate or humic acid types are more easily assimilated by the plant.

The popular EDTA chelated trace minerals don't appear to be as effective as other chelates, especially when foliar fed, but will generally test better than sulfates and oxides. One of the most common sources of trace minerals used by alternative ag adherents is seaweed. Seaweed contains very small amounts of all trace minerals found in the ocean or lake in which it grew. Purchased in either a dry soluble or liquid form, these are excellent products to supplement fertilizer applications. As with everything else, energy readings and plant or soil benefits vary depending upon form and manufacturing source.

Boron B — Boron functions as a regulator in the plant's metabolism of carbohydrates and hormones. Hollow hearts in vegetables have generally been associated with boron shortages. Alfalfa has been identified as particularly needing boron. Generally, boron is not available in high pH soil or soil low in organic matter. Boron can be obtained from using 20 Mule Team Borax soap powder; in a powdered form under the trade name of Solubor, and in other chelated formulations. Energy levels will vary with source.

Chlorine Cl — Chlorine is an essential growth element. It is usually adequately provided naturally from thunderstorms.

Cobalt Co — Cobalt is associated with plant enzymes and protein synthesis and is needed for fixing nitrogen by bacteria. Energy levels will vary depending on the source.

Copper Cu — Copper is largely associated with plant enzymes. It regulates plant bark "stretchiness." It may be somewhat immobile in higher pH soils. Copper is known for its fungicidal qualities. Energy values will vary depending upon the source.

Iron Fe — Iron is needed for chlorophyll and protein synthesis, photosynthesis, is a component of enzymes, and plays a major role in the oxygen-carrying system. Energy varies depending on the source.

Magnesium Mg — See Magnesium above.

Manganese Mn — Manganese is essential for the activity of many enzyme systems within a plant. Chlorophyll production and carbohydrate and nitrogen metabolism all depend upon this essential micronutrient. Manganese is important to both crop and animal nutrition. It is crucial to reproductive functioning, is a relatively heavy metal, and requires good biological activity and soil fertility levels to make it available to plants. Tubers with sunken "eyes", meatless nuts, stone-fruit pits that split, or seeds which fail to germinate are associated with depressed levels of manganese. Energy levels will vary.

Molybdenum Mo — Molybdenum is needed for fixing nitrogen in legumes and for the reduction of nitrates in the formation of protein. It is critical for "grasses" and other crops requiring very little potash. It can be toxic in high doses and must be labeled with a "caution" label in concentrated forms.

Sulfur S — See Sulfur above.

Zinc Zn — All plants need zinc for normal growth and reproduction. It is an essential part of certain enzyme systems related to plant growth. Energy values will vary.

Common Nutrient Symbols which Apply to Farming:

Aluminum . . . . . Al	Arsenic . . . . . . As	Barium . . . . . . Ba
Boron . . . . B	Bromine . . . . . Br	Calcium . . . . . Ca
Carbon C	Chlorine Cl	Chromium . . . Cr
Cobalt . . . . Co	Copper . . . . . Cu	Fluorine . . . . . . F
Hydrogen . . . . . . H	Iodine I	Iron . . . . Fe
Lead . . . . Pb	Manganese . . Mn	Magnesium . . Mg
Molybdenum . . Mo	Selenium . . . . . Se	Nitrogen . . . . . . N
Oxygen . . . . O	Phosphorus . . . P	Potassium . . . . . K
Silicon . . . . Si	Sodium . . . . . Na	Sulfur . . . . S
Zinc . . . . Zn		

Chapter 12
pH, Lime & Calcium

The misunderstanding and misuse of the pH, lime and calcium concept has cost farmers millions of dollars. The damage done is probably second only to the damage seen from using the acid/salt/anhydrous fertilizer concept. The general concept taught is that pH is an important factor in soil. The pH is raised or lowered in an attempt to make nutrients available or adjust soil conditions to those recognized as being preferred by various types of plants. Lime is applied to neutralize soil acids by raising the pH.

pH Explained

Let's take a look at how Dr. Reams viewed pH. In a water solution, there is a supply of both hydrogen (H^+) ions and hydroxide (OH^-) ions which comes from the natural separation of water molecules, usually written as H_2O but more accurately written as HOH (to help our understanding of how you get H^+ and OH^- ions). In a neutral solution with a pH of 7, the amount of H^+'s and OH^-'s is considered virtually equal. By convention established a long way back, science measures the number of H^+ ions only and calls the mathematically derived number the pH. As the H^+ ions increase, we consider this an acid condition. However, due to the formula and method of calculating, the numerical value of the pH goes down (below 7) as the hydrogen ions (H^+) increase. As the number of hydrogen ions goes up, the number of hydroxyl ions (alkaline or base condition) goes down correspondingly. When you add a base (KOH or K^+ and OH^-) to an acid solution, the balance shifts the other way, and the pH goes up.

In agriculture, we don't usually work with strong acids like sulfuric acid or strong bases like calcium hydroxide. Instead, we work with products that aren't as strong so as to prevent soil and microbial damage. The products we add are able to make mild acids or bases after they have become mixed with the soil solution. For instance, when we add epsom salt ($MgSO_4$), we can get SO_4^- ions which tend to separate H+ from the water. The H^+ tend to stay available, in association with the $SO_4^=$ (forming sulfuric acid), while the OH^-'s tend to more tightly bind with the magnesium (Mg^+) or other positive ions. The net result is more H^+'s or lower pH which is the expected result from adding sulfur to soil. When we add limestone, or calcium carbonate, we get (Ca^{++} and $CO_3^=$). In this case, we get the carbonate, $CO_3^=$, reacting with the excess H^+'s to neutralize them into carbon dioxide and water. In summary, soil pH is manipulated upward or downward through the addition of substances like lime (base) or sulfur (acid) to the soil.

However, all the chemistry aside, Dr. Reams felt the concept was flawed. He argued that in a laboratory situation, the idea of measuring the absolute amount of H^+'s in a water solution provided a workable concept (pH). But when you work with a living solution, as in human blood or a soil solution, he felt the other materials in the blood or soil couldn't possibly allow you to measure the actual amount of H^+ ions with the same meter accuracy because now you have an interference factor. Now the H^+'s have to fight their way across the space between the pH meter electrodes and this effect would alter how many H^+'s were measured. It's like saying you can run the length of a football field just as fast dodging all the players on it as you can doing a straight 100-yard dash.

pH as Resistance

Dr. Reams viewed pH as a measure of resistance rather than of total/actual H^+ ion concentration. If the soil pH is low, e.g., 4 to 5, you have less resistance and the pH meter reads a higher concentration of H^+ or acid. When the pH is high, e.g., 8 to 9, you have more resistance to flow and the meter registers low H^+ concentrations. This understanding can help explain phenomenon such as manganese toxicity at low pH's. The resistance is low due to the flow of positive ions like hydrogen (H^+) and manganese (Mn^{++}). Manganese is highly paramagnetic (attracted to a magnet) and will flow rapidly on the earth's magnetic field. With the reduced resistance, it can overdose a plant.

Conversely, iron tie-ups can occur at high pH's when there is great resistance to flow.

As the chart on the following page shows, nutrient (mineral) excess and deficiencies are related to pH.

The primary purpose of liming, according to conventional wisdom, is to neutralize soil acids so the pH will change to a level where the plant will grow efficiently. The soil acids are presumed to result from the cation exchange concept whereby roots give off H's in exchange for other positively charged nutrients from off the soil colloid such as calcium (Ca^{++}), magnesium (Mg^{++}), and potassium (K^+). The net result is to build up H's on the clay colloid which lowers the pH and makes an acid soil. Heavy emphasis is placed on pH which supposedly allows you to add the right amount of lime and everything will be just fine.

pH Stability

Besides the resistance versus the absolute amount of H's controversy, there is also the issue of pH stability. Canadian government officials in western provinces ran a two-year study of soil pH by checking readings every two hours, twenty-four hours a day for two years. They found the pH made wide (24 to 48 hour) swings on both sides of a neutral pH 7. In other words, pH is a dynamic concept.

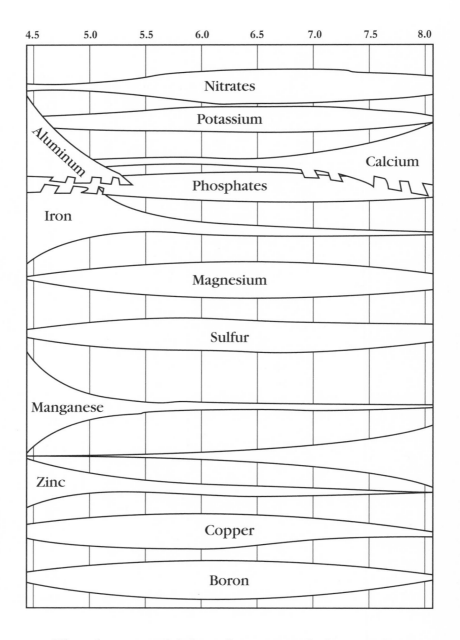

The relative availability of 12 essential plant nutrients in well-drained mineral soils in temperate regions in relation to soil pH. (Source: Liming Acid Soils, Leaflet AGR-19, 1978, University of Kentucky.)

According to Dr. Reams, without pH changes the necessary biological and chemical reactions required in the soil would probably not take place since many chemical reactions and biological responses are subject to the immediate environmental conditions of pH, temperature, moisture, etc. These biological responses will, in turn, affect the immediate environment. It is better to view pH as something that will take care of itself if major imbalances are corrected with appropriate inputs designed to bring the soil into a proper range. By concentrating on the balancing of nutrients and the stimulation of biological activity, you will allow the pH to "do its thing" to the betterment of the living soil.

In some cases, the type of soil precludes making major pH changes because of cost. For example, high-CEC soils are difficult to alter inexpensively. In these cases, you can work with wherever the soil is. The main problem that has developed over the last 30 years is the general rising of pH's from several factors other than adding high-calcium lime. Dolomite has raised pH with the accompaniment of excessive magnesium; high potash (mined or manure) applications can raise pH; and anaerobic secretions can also raise pH. This was accompanied by the advice that you "shouldn't lime if your pH is in the high 6s or 7s." Why can this be a problem?

Soil pH will rise from adding a liming material like calcium carbonate, calcium oxide, or calcium hydroxide. But pH will also rise if any positive ion is added. The major positive ions that attach themselves to the negative clay colloids of your soil are calcium, magnesium, potassium and sodium. If you don't differentiate between ions and simply consider pH, you are falling into the pH trap and you may have imbalanced nutrients, particularly a shortage of calcium. Since the available calcium determines the total yield of your crop, you could be losing yield and test weight from being caught in the trap. So the first rule is: *calcium is king* and the second rule is: *don't use pH to determine if you need to apply calcium.*

There are several ways to determine if your soil needs calcium. The obvious way is by soil testing — a CEC test is the standard method used. A CEC test will give you the percentage of base saturation of the above-mentioned positive ions and then you can evaluate your calcium levels. Standard wisdom calls for 65-70 percent of your colloid cations to be calcium. The LaMotte or available soil test and electronic scanner's energy soil test methods can also be used to measure calcium levels, but we prefer the clear picture of a CEC test. The scanner test can be used to determine appropriate types, amounts, and times of calcium application.

Lime Sources

What lime do I use? The first choice, in most situations, would be a fine grind of a high-calcium lime with as little magnesium as possible. A 5 percent magnesium is great, 10 percent is getting questionable and above 10 percent is grounds to check out another source. Once you have above a 10 percent magnesium content, it should be called dolomite or dolomitic lime. This is not a recommended form of lime — even though it is widely promoted. Dr. Reams suggested you avoid dolomite for three reasons. The most impressive one has to do with nitrogen release. Magnesium is antagonistic to nitrogen as seen in the use of epsom salts as a treatment for nitrate poisoning in cattle or an epsom salt spray on fruit trees to stop apple drop due to nitrate-weakened stems. When the magnesium releases from dolomite, it can cause nitrogen to release as a gas.

The second reason for avoiding dolomite has to do with the physical structure of soil which can affect both chemical and biological processes. The high amounts of magnesium applied as dolomite tends to overload the clay colloids once the magnesium ions are released. As the percentage of base saturation of magnesium rises, the potential for soil stickiness increases. If mud builds up on boots, tires, etc., and then turns hard and compact in dry weather, you probably have a magnesium problem.

The third reason has to do with forages. If forages grown on a high-magnesium field reflect the same overages/imbalances as high-magnesium soil, this can cause problems, particularly with ruminant livestock.

Any one of the above three problems are sufficient reason to avoid dolomite as a general practice. High-calcium lime can be applied at most any pH, but is usually reserved for a pH of 7 or below. When the pH is above 7, gypsum (calcium sulfate) is preferred. This assumes that the high pH is due to sodium, magnesium or potassium. If the soil really is calcitic (very high in calcium), then the additions of sulfur forms other than gypsum would be best. Sulfur could be applied as dilute sulfuric acid, thiosul, NKS Plus or ammonium sulfate. There are many other sources of limes that can be used. Native deposits of marl (lower calcium content, chalk-type deposits), calcium oxide deposits or by-products, mine tailings and cement dusts may be readily available at reasonable costs. Get on the phone and check it out.

Because lime is added in large amounts compared to other nutrients, it is wise to temper your applications with an eye to the soil and your pocketbook. We prefer not to add more than a ton of lime in the spring to avoid possible root dehydration and major energy, chemical or physical changes. Two or more tons may be applied in the fall. There are many cases where three or more tons are used with seemingly good results. Is it possible to over lime? Yes, especially if you ignore the tie-up effects liming can have on trace minerals.

More recently, adding lime at as little as 300-400 pounds per acre has proved to be beneficial. In other words, the calcium in the lime is being used as a nutrient. Cases have been observed where the adding of high-calcium lime to a soil with a pH well over 7 has dropped the pH to below 7. All is not as it seems in the complex world of living soil.

Chapter 13
Moon Cycles
& Plant Manipulations

Nature and Moon Cycles

Nature has her own growth timetable which naturally cues the plant regarding growth and fruiting cycles. Some of the variables considered important include the length of sunlight hours, the amount of heat received each day, the genetics of the plants involved, rainfall, and the soil fertility balance. Above all else, nature wants to continue the species through reproduction of seed. Many farmers have seen this happen, for instance, when a summer drought threatens crops. Grains will suddenly head out even though they may only have reached 15 inches in height. Nature is interested in the preservation of the species, and this is her method of preserving life through seed production.

The moon cycle also affects the growth cycle. On a full moon, the ground will be warmer while the air will be cooler. This reverses itself on a new moon. A full moon represents a stronger gravitational tug against the earth causing high tides and also causing the soil to be "looser." Many a farmer has recognized this while setting fence. On one occasion, he digs the hole, sets the post and still has soil left over while on another occasion he needs additional soil to fill the hole. Why? The soil will loosen or compact depending upon the phase of the moon. As the moon changes phases, the temperature difference between the soil and the air creates new energy in the soil. Planting by the moon phases will help crops to respond to these naturally occurring energy changes.

Plant Manipulations

Farmers have witnessed corn stalks which grew only three or four feet high, tasseled out and began setting ears. They have also grown corn stalks twelve feet high which never set an ear. The same effects have been observed for grains. Why is this possible when genetics would dictate otherwise?

Fertilizer Effects on Plant Growth and Fruiting

Basically, there are two types of responses to fertilizer energies — plant growth and plant reproduction. Understanding this important concept will assist the farmer in properly controlling his fertility (energy) program by allowing him to better time and control his energy release. These concepts can also allow the farmer to push crops in a given direction in order to assist nature or to counter nature's direction.

Fertilizers Producing Growth Responses

Fertilizers producing growth are generally considered to be "sweet." The farmer knows calcium (lime) is considered sweet. He uses lime to sweeten soil soured by acid fertilizers and nitrogen breakdown. Other minerals should also be considered sweet. These minerals produce growth response in plants. The basic growth-producing (sweet) minerals are: calcium, potassium, chlorine and nitrate nitrogen. Each of the above nutrients are necessary, in the correct balance, for proper plant growth. The key word here is balance.

Calcium is the major mineral needed to grow strong, healthy plants which naturally resist insect attack. Calcium provides crops with good mineral balance and provides soil microbes with the proper conditions for their proliferation. Calcium is needed more by weight and volume than any other mineral nutrient.

Chlorine is adequately supplied each year through naturally-occurring thunderstorms. It is not usually found to be a limiting factor by its absence. Its addition in the form of potash, 0-0-60, does not need to be made.

Farmers know nitrogen can cause growth, and so they tend to apply it excessively. Excess nitrogen, however, will cause plants to hold excess water and excess nitrates. Plants with excessive nitrogen may under-produce such essentials as vitamin C. This imbalanced condition can set the stage for disease and insect attack. The excess water can cause early spoilage while the excess nitrates can cause storage and animal health problems. Applications of over forty pounds of nitrogen per acre also shuts down naturally-occurring microbial life intended to extract nitrogen from organic matter in the soil.

Growth energies should be emphasized during seed germination and early growth stages. Farmers usually apply growth fertilizers preplant and at planting. These may be foliar sprayed until the 40th to 45th day after plant emergence, but watch the crop, not the calendar.

Fertilizers which contain only growth energies are:
calcium nitrate . . . . .excellent
potassium nitrate . . . . .excellent
potassium chloride . . . . .avoid

Fertilizers Producing Reproduction or Fruiting Responses

Fertilizers producing fruiting responses are considered to be *sour* and, consequently, will sour or acidify the soil as well. The basic fruiting minerals are: phosphorus, sulfate sulfur, manganese and ammonium nitrogen.

Fruiting energies are normally applied when the plant starts to set fruit buds. Buds indicate the plant is starting to change from growth to fruit production. The farmer can influence (speed up or slow down) this process through his fertility (energy) program. However, with many plants, reproductive processes may begin before any external signs are visible. An electronic scanner or close visual inspection may be very useful in determining the early changes (15-25 days after emergence) and allow you to influence them with foliar or sidedressing materials. Foliar sprays and side dressing, applied about 40-45 days after emergence on corn, give the added reproductive energies needed to develop full ears and increase the potential for filling out second or third ears. By planting by the moon schedule, this naturally occurring progression is enhanced.

Combinations of Growth- & Fruit-Producing Fertilizers

Phosphorus is the primary plant nutrient needed to combine with other nutrients entering the plant. This combining results in a new electromagnetic frequency for the molecule which allows the mineral attached or associated with the phosphorus to be attracted to the correct place in the plant. The combinations produce calcium phosphate, manganese phosphate, etc. which directly affect the correct utilization of the mineral by the plant.

This results in optimum efficiency which leads to increased immunity, nutrition, production, and high brix (sugars). Plants that are high in balanced minerals will taste better, be more filling and satisfying, help eliminate compulsive eating patterns (since mineral needs will have been satisfied), and will store better.

Sulfur, required in the sulfate form, is essential to the soil. It is a strong energy source which not only pushes plants to reproduce, but also assists in releasing calcium. It is required for the production of amino acids in the formation of protein.

Manganese is also considered a fruiting element. Dr. Reams called it the "seed of life." No reproduction takes place without adequate manganese in plants, livestock and humans. It is a very heavy element and, consequently, is difficult for plants to take up. This is especially true for plants in a weakened condition due to stress or inadequate mineral balance. Seed lacking in manganese will often rot in the soil rather than sprout. A balanced soil will produce a strong energy pattern which assists manganese in entering the plant.

Ammonium nitrogen is the nitrogen form required for fruiting and is held in the humus (carbon) structure of soil. Current soil tests show a dramatic inability of soils to hold ammonium nitrogen even though anhydrous ammonia is usually applied at planting. Soil microbes that are in low humus conditions will transform ammonium nitrogen into the nitrate form of nitrogen which can result in considerable energy loss. Release of excess ammonium is a major cause for the intensification of plant signals which beckon insect attack. This happens because the soil, lacking active carbon, cannot hold the ammonium. By applying anhydrous preplant, the farmer not only forces soil microbes to change the nitrogen into a form (nitrate) the plants can use for growth, but he also loses much of the ammonium energy by the time the plants require it for reproduction.

Common fruit-producing fertilizers:

ammonium sulfate	excellent
11-52-0	good to excellent
superphosphate	good to excellent
triple super phosphate	not recommended
10-34-0	good
ammonium thiosulfate	excellent
hard rock phosphate	fair to good
colloidal phosphate	good to excellent

Combinations of growth- & fruit-producing fertilizers:

ammonium nitrate	good to excellent
calcium sulfate	good to excellent
potassium sulfate	good to excellent
sul-po-mag	good to excellent
colloidal phosphate	excellent
9-18-9	fair to good to excellent
3-18-18	fair to good to excellent
10-20-10	fair to good to excellent

Farming Applications

To be in proper balance, soil must always contain a mixture of both growth and fruiting energies. Once that base is established, the soil or plant can be manipulated through the addition of one form or the other. In planning fertility programs, the above information can play a crucial role. Generally speaking, apply the growth producers in the spring and the fruit producers about 35 to 45 days following crop emergence.

When farmers couple this concept with planting by moon phases, they accentuate the effect which occurs in nature. Foliar feeding of plants is an especially powerful tool when using these concepts. The farmer can assist a crop through foliar feeding and enhance the basic plant response (growth or fruiting) he desires. Since plants respond from eight to twenty times more efficiently to the energies applied through foliar feeding compared to ground applied, the farmer is maximizing his dollar.

Crops which are harvested for their leaves, for instance — clover, hay, spinach, cabbage, and herbs — would be fertilized with additional growth-producing fertilizers. Crops harvested for their seed, such as corn, wheat, tomatoes, and peppers, would require both growth- and fruit-producing fertilizers with the timing important — varying with the length of the growing season.

A rough estimate for determining when to change to reproductive energies is to watch for when the earliest plants first begin to set buds. If the budding process has begun, the soil and plant have already changed and a foliar or sidedress application may greatly assist enhancing the fruiting response of the entire crop.

Using these concepts, the farmer can direct growth patterns in the desired directions. Using the LaMotte soil test he can monitor the monthly or bimonthly nutrient levels in his fields and take corrective actions where needed. Using the electronic scanner, he can often obtain similar results. Fruit trees and perennial berries are generally considered separate from field or row crops. Fruit trees usually begin preparing for the next year's fruiting process in mid- to late summer. This is also the time the current fruit is growing in size. Trees produce their blooms in early spring and fruit immediately after.

Thus, the farmer would apply fruit-producing foliar fertilizers in the spring and until mid-summer, at which time he would switch to growth producers to build the size of the fruit and new growth for the next season. Many growers do not provide adequate energy at this stage with the resultant low budding, blossoming and fruit set the next year. Dry corrective/balancing fertilizers may be added from late summer until snow fly.

Fertilizers also serve to attract similar nutrients to themselves. The nutrient energies being attracted in the soil are carried over the surface of the earth on the same magnetic currents that operate a compass. To maximize the attraction of energies, the farmer must consciously plan to create as broad a "magnetic base" as possible. Raising organic (humus) levels, balancing calcium, magnesium and phosphate levels is required here. This creates a highly beneficial environment for soil microbes which will add to soil fertility.

A fertile soil in good tilth will continue to grow more fertile because the balance of nutrient energies serves to attract balanced energies in return. This works in the opposite as well. By adding the toxic energies with "kill" frequencies to the soil, the soil will attract similar energies to it. As the soil becomes more and more out of balance, it will attract an imbalance of energies to it. The rich get richer, the poor get poorer.

Keeping all these concepts in mind, the farmer is in a position to make definite changes in his soil environment. He should not get discouraged if the process takes a two or three year period since this would be a rather short time to turn a soil completely around. After years of abuse, the soil will need time to recover.

Chapter 14

Refractometer

What It Is

A refractometer is a simple hand-held instrument which is used to measure a variety of solutions from cutting oils to soups to acids to sugars. Actually, the refractometer measures dissolved solids. In dealing with plant juices, the majority of solids are compounds manufactured from carbon, hydrogen and oxygen. These compounds are primarily carbohydrates (starches and sugars). The refractometer consists of a tube-shaped instrument with an adjustable eyepiece at one end and a prism with a plastic cover at the other end. The light passing through the liquid being tested is refracted (bent) in relation to the amount of dissolved solids found in the liquid.

It has been said that the refractometer measures the health of the plant. The readings obtained from measuring plant juices can be compared with the chart at the end of this chapter. To obtain readings in the excellent range or higher requires the plant to be in good condition from a balanced mineral perspective. Since plants grow from minerals or mineral energies, it is these minerals (or lack of them) which is being reflected in the refractometer reading of the plant juices.

To bring plants from consistent low readings to consistent high readings usually requires a year or two of intensive building with accompanying foliar sprays. It can be done, though, as demonstrated by the experience of many farmers across the United States. Basic soil fertilizing is the first step. Using only quality, high energy, bio-enhancing products is the second step. Foliar spray assistance is the third step.

How It Is Used

The farmer squeezes juice from the plant's leaves, stem or fruit onto the prism and gently lowers the plastic cover to cause a layer of plant juice to cover the prism. He then holds it to his eye looking in the direction of a light source and observes where the blue and white backgrounds come together. In the middle of the circular field is a series of numbers starting from 0 at the bottom and, for agricultural models, ending at 30 or 32. By observing at which number the blue and white fields meet, the farmer can read the brix reading.

It is important for the farmer to take readings and keep records of the results. By comparing data from field to field and year to year, he is able to better assess his progress toward fertility goals. Weekly readings are most helpful. Take readings at the same time of day and from the same area of the plant. For instance, pick the second or third leaf level of a corn plant and stay with that level throughout the season. Additional leaf levels can be measured as well. Keep in mind that changes in relative moisture levels can change the dilution of the juices and influence brix readings.

One of the most challenging aspects of using the refractometer is, "How do I get the juice out?" This question is usually answered by using a garlic press. Many garlic presses, however, are not constructed strong enough to withstand the pressure often needed to press out the plant juices. You may want to modify a pair of pliers or other instrument to crush the stem or leaf. As a last resort, equal volumes of distilled water and tissue can be blended. The reading should be multiplied by two to adjust for the dilution.

Interpreting the Readings

Refractometer reading variations will occur due to several variables. Testing during different hours of the day will affect the reading. Storms or other impending weather changes will lower the reading. Sometimes a drought will raise the reading since the water content is low and the juices are more concentrated.

Although the reading should remain constant throughout the length of a corn plant, it is not uncommon to find the readings varying greatly. This may indicate a lack of nutrition or energy sufficient to keep the levels equal. It is generally held that a clear, distinct line separating the blue and white fields indicates a more acid condition while a fuzzy line indicates better calcium levels and a more alkaline condition.

Refractometers are usually calibrated using distilled water to read "0" at 20 degrees Celsius. Correction tables reveal how much the reading will vary at different temperatures. These differences, for our purposes, are not worth considering. If a plant reads 4, it means a "poor" reading regardless that the true, or correct, reading would be 4.23.

By raising the sugar levels (brix) of his plants, the farmer is able to feed increasingly more valuable feedstuffs. He will need less purchased mineral supplements, his livestock will become more content on less feed, he will have raised the energy value of his feeds with resultant improvements in animal performance, and his veterinary bills will reduce substantially. All of these results have been documented by farmers using biological methods.

The following chart gives a range of refractive values for commonly grown fruits and vegetables. Compare these values with your crop readings and develop a fertility program to systematically raise your values.

Crops may test below the poor or above the excellent readings. Within a given species of plant, the crop with the higher refractive index will have a higher sugar content, higher mineral

Refractive Index of Crop Juices
Calibrated in 0 Brix

Fruits	Poor	Excellent
Apples	6	18
Avocadoes	4	10
Bananas	8	14
Cantaloupe	8	16
Casaba	8	14
Cherries, sweet	6	26
Cherries, tart	6	18
Coconut	8	14
Grapes	8	24
Grapefruit	6	18
Honeydew	8	14
Kumquat	4	10
Lemons	4	12
Limes	4	12

Vegetables	Poor	Excellent
Asparagus	2	8
Beets	2	12
Bell Peppers	4	12
Broccoli	6	12
Cabbage	6	12
Carrots	4	18
Cauliflower	4	10
Celery	4	12
Corn Stalks	4	20
Field Corn	6	18
Sweet Corn	6	24
Cow Peas	4	12
Endive	4	10
English Peas	4	10

Mangos 4 14
Oranges 6 20
Papayas 6 22
Peaches 6 18
Pears 6 14
Pineapple 12 22
Raisins 60 80
Raspberries 6 14
Strawberries 61 6
Watermelon 8 16

Grasses

Alfalfa 4 22
Cotton 4 22
Grains 6 18
Rice 4 16
Soybeans 4 16
Turnips 4 10

Escarole 4 10
Field Peas 4 12
Green Beans 4 10
Hot Peppers 4 10
Kohlrabi 6 12
Lettuce 4 10
Onions 41 0
Parsley 4 10
Peanuts 4 10
Potatoes, Irish . . . 3 7

Potatoes, Red 3 7

Romaine 4 10
Rutabagas 4 12
Squash 6 12
Sorghum 6 30
Sweet Potatoes . . . 6 14
Tomatoes 4 12

content, higher true protein content, and a greater specific gravity or density. This adds up to a sweeter tasting, more minerally nutritious food with a lower nitrate and water content and better storage characteristics. It will produce more alcohol from fermented sugars and be more resistant to insects, resulting in a decreased insecticide usage. Crops with a higher sugar content will have a lower freezing point and be less prone to frost damage.

Soil fertility needs may also be ascertained from this brix reading. Plants or soils that do not produce significant brix do not contain sufficient biologically active phosphorus or calcium. The soils will also reflect the same shortages by the presence of broadleaf weeds and sour grasses.

Chapter 15
Foliar Feeding

What Is Foliar Feeding?

Foliar feeding is a highly efficient method of providing needed nutrients to crops. Research conducted by Dr. Silvan Witwer, Michigan State University, in cooperation with the Atomic Energy Commission in the 1940s, found plants to utilize foliar-fed nutrients anywhere from eight to twenty times more efficiently than those applied to the soil. Their research concluded that trees benefited from and absorbed foliar fed nutrients even during mid-winter months.

Foliar fed fertilizers seem to bypass problems associated with root absorption, such as nutrient competition, nutrient tie-ups, leaching, and soil interactions. Foliar uptake requires the same light, temperature and oxygen variables as does root uptake. It may have a significant effect on lower CEC soils. The mobility of different nutrients once in the leaf varies widely. Whereas all nutrients are initially absorbed extremely rapidly, nitrogen, phosphorus, potassium, copper, manganese and zinc are readily translocated. Calcium, boron, iron, magnesium and molybdenum tend to remain in the leaf after they are absorbed and have little tendency to translocate.

Foliar feeding works best in cooperation with a good soil fertility program. Without the proper soil fertility base to begin with, foliar spraying can have very mixed results. Good success could be obtained in certain instances where the right nutrients were sprayed at the right time for the growing plant, yet these would be exceptions to the rule. One would expect to see minimal positive results if good fertility programs were not followed. Foliar feeding is not as efficient when it is used as a rescue procedure.

Foliar feeding is intended to strengthen basic fertility (energy) programs. It is used to help swing the plant from growth to fruiting, to alleviate a stressful situation, to counter leaching brought about from steady rains, to give an added push, and to keep the plant's energy at optimal levels.

How to Foliar Feed

Foliar feeding can be done using a typical boom sprayer with 10- to 20-gallon nozzles. If possible, it is best to use nozzles which will produce a mist or fog. Plants feed mainly from the under side of leaves through openings called *stomata*. The plant leaf hairs surrounding the stomata will attract nutrients within fine water droplets. It is possible to purchase nozzles which produce a cone-shaped spray pattern and which also spin the spray upon leaving the nozzle. These have been found to be highly effective.

A new generation of mist sprayers is available which will produce a spray mist and blow it 40 or more feet across the field. These sprayers are actually the most cost effective to use because, contrary to logic, the finer the water droplet coming from the sprayer, the more dilute the spray solution can be and still accomplish the feeding task. Mist blowers will work effectively using only one-third or less the amount of fertilizers needed for field or boom sprayers.

Vegetable and orchard growers have long used foliar spraying, however, the major purpose has been to apply chemicals. It's not unusual for these sprays to be used every one to two weeks. Although most farmers spray toxic chemicals for blight, fungus or insect control, it is obvious that supplementing these toxic sprays with a good nutritional "diet" could be very effective. In fact, many farmers now know that periodic nutritional foliar sprays can not only *save* a crop but can also *make* a crop.

When grain is in its early development, for example, it is possible to slice open the growing stem and see, through a ten-power lens, the grains of oats or wheat developing on a head. Prior to their development, a fruiting foliar spray could result in a larger number of seeds being set. After this critical development period has passed, it is no longer possible to influence seed head development in terms of a larger head formation. Now it is only possible to influence how large the seeds will form and how heavy (test weight) they will be. Once the head has formed, foliar sprays will definitely assist in producing larger, heavier grains.

Wheat growers in the Lake Palouse area of Washington have grown 200 bushels of wheat using good fertility and foliar spraying. Their wheat is so good that the farmer can slice a kernel from top to bottom, dividing it in half, and grow two plants from the now split seed.

Spray pH

Foliar sprays are especially important when considering the growth or fruiting concepts presented in Chapter 13. By measuring and manipulating the solution pH, the farmer can direct his crops in the intended direction for growth or fruit production. If the spray solution has a pH lower than 6 and the farmer wants growth energies, he should consider adding baking soda to raise the solution pH over the 7 level. If on the other hand fruiting energies are desired, he would lower the solution to read below 7 by adding apple cider vinegar. Note: If spraying a chemical or nutrient that is pH sensitive (subject to alkaline hydrolysis), you may have to keep the pH at 6 or below to prevent loss of efficacy of the product.

When to Foliar Feed

Like everything else, timing of foliar spraying is important. Foliar sprays are extremely beneficial to young plants in the third to sixth leaf stage and especially during the 35 to 45 days after emergence to change the plant from growth to fruiting. Young foliage seems to take up nutrients more readily than old. Foliar sprays are beneficial for stress caused by transplanting, hail, wind, heat and drought. They are effective at fruit bud formation, after petal fall, and during pod or fruit filling. A foliar spray is indicated whenever the refractometer reading drops two or more points from nutritional shortages.

It is best to foliar spray in the mornings and evenings when relative humidity is high and generally not in the middle of the day. When temperatures soar, the effectiveness of the spray drops considerably. When using a boom sprayer, use high pressure and purchase atomizer nozzles if possible. Tip standard nozzles back about 90 degrees so the spray will roll up under the leaves. Keep active ingredients on the dilute side, e.g., 1 to 2 quarts per acre for majors and a few ounces for traces. It may be possible to use as little as 2 cups of active ingredient per acre and still be effective, especially when using a mist blower. The use of a wetting agent will often assist the solution to break down and homogenize.

Many farmers are experimenting with adding food grade 35 percent hydrogen peroxide to their sprays. Used at the rate of 2 to 6 ounces per acre, this often has a very beneficial effect to both the plant and the soil. Always add hydrogen peroxide to the spray tank before adding any other chemical or fertilizer.

Predetermining Foliar Spray Effectiveness

Successful farmers and consultants have found a simple method of "asking the plant" if a fertilizer spray is desirable, neutral or undesirable. The procedure simply entails checking the crop refractometer reading before and then 30 to 45 minutes after the fertilizer spray in question has been applied. Only a small test area is evaluated, thereby eliminating the need to spray an entire field only to learn that the spray used was ineffective or even detrimental.

Testing Procedure

Construct two or more plastic or wire rings that encircle an area of 5 or 10 square feet. (The area of a circle is approximately 3.1 (actually π = 3.1417...) times the radius squared ($A = \pi r^2$) and the circumference of a circle is approximately 2 X 3.1 X radius ($C = 2\pi r$). A ring encircling 5 square feet would have a diameter of 2.5 feet (30 inches) and a circumference of 7.9 feet (95 inches). A ring encircling 10 square feet would have a diameter of 3.6 feet (42.8 inches) and a circumference of 11.2 feet (134.5 inches). A 10-square-foot ring makes calculations easier, but is more difficult to carry around unless you make it easy to disassemble. Since there are 43,560 square feet in an acre, 10 square feet equals 1/4,356th of an acre.

Take these rings to the field and drop them 3 to 5 feet apart. One ring will be used as a control or check area while the others will be designated as tests. Check the refractometer reading within each ring and record these readings. In a quart spray bottle, mix the exact fertilizer and water ratio that your sprayer is calibrated to apply. For example, if you would apply 2 quarts per acre of 6-12-6 in 20 gallons of water, you would mix 0.8 ounces of 6-12-6 in 1 quart of water. Twenty gallons equals 80 quarts, so a 1-quart mix for your spray bottle would be a 1/80th portion of your acre mix. This is calculated by multiplying 2 quarts by 32 fluid ounces per quart to get 64 ounces. Divide 64 by 80 to get 0.8 ounces of 6-12-6 for one quart of water (0.8 ounces = 1.6 tablespoons = 1 tablespoon + 2 teaspoons).

Next, you need to determine the number of squirts from your spray bottle to apply to your 10 square foot test area in order to equal a spray rate of 20 gallons per acre. Since 10 square feet equals 1/4,356th of an acre, you will need to apply 1/4,356th of 20 gallons or approximately 0.6 ounces, which is about 2.4 tablespoons. This is calculated by multiplying 20 gallons times 128 fluid ounces per gallon (2,560 ounces), and dividing this number by 4356. Since there are about 2 tablespoons per ounce, 0.6 ounces times 2 tablespoons per ounce equals 1.2 tablespoons or 1 tablespoon plus 2/3 of a teaspoon.

Take a measuring cup and count the number of squirts from your 1-quart spray bottle to equal 0.6 ounces or 1.2 tablespoons. This is the number of squirts you will apply to your 10-square-foot test area to equal the amount that would be applied if you sprayed the field with your sprayer at 20 gallons per acre. After misting this spray mix on the test area, wait 30 to 45 minutes and recheck the refractometer reading of the crop in the test area as well as the control. If the test area brix reading increased by at least one full point, the spray is desirable and would benefit the field. You can then spray the entire field with confidence that the spray is beneficial. You can mix several different sprays and check each. If the refractometer reading remains unchanged or drops, the spray is undesirable for that particular day for that particular crop. It is not necessarily a reflection on the product quality itself, but rather the plant's need at that time.

The only exception to this guideline is where there might be a delayed reaction where no change in refractometer reading is observed for several to 24 hours. In this case, leave the rings in the field for 24 hours and recheck the brix reading of both the control and the test areas. If an increase in brix reading of the test area is observed, this spray would then be sprayed on the entire field. The delayed response may occur where a specific prescription has been formulated for a particular field and the spray test is a verification of its appropriateness.

In determining how much active material to mix in the spray bottles, use the following guidelines. Assuming a 20-gallon per acre spray rate, mixing 1 quart of solution for a spray bottle would require the following amounts of active material to be placed in a one quart container and the filled to the one quart level with water:

Amount of Active Material/Acre
& Equivalent Active Material/Quart:

1 pint	=	.2 oz. (1.2 tsp.)
1 quart	=	.4 oz. (2.4 tsp.)
2 quarts	=	.8 oz. (1.6 Tbsp.)

(1 oz. = 2 Tablespoons = 6 teaspoons)

It is time growers become aware that nutrition plays the primary role in plant health and seed development. It is possible to greatly influence crop production by using these ideas. It's possible, perhaps probable, that the amount per acre applied is not as important as the substance, or more accurately the "energy" of the substance applied. Within reason, the authors would guess that for foliar feeding even doubling the application rate from 1 to 2 quarts would not make a substantial crop response difference. It's the "energy" being applied and the clean water as a carrier which is more important. Having gone through the above calculations, a general conclusion could be made by simply making fertilizer solutions containing 1 tablespoon active ingredient per quart of water and applying.

Chapter 16
The Role of Nutrition

Nutrition Based Upon Minerals

Dr. Reams considered such problems as insect damage, disease, and the inability to yield up to or near genetic potentials to be caused by mineral (or nutrient) deficiencies. Minerals are the foundation of life. Minerals provide the building blocks from which the soil or body can produce the other requirements necessary for proper growth and functioning. He understood the role of vitamins and enzymes in living systems and knew that they could or should be provided either from the foods eaten or from the body producing them from other building blocks. Amino acids, for instance, are the building blocks of protein. Many amino acids need sulfur in their chemical makeup. When these elements are present in the proper ratio, the organism, be it microbial, plant, animal, or human, can function according to its maximum genetic potential.

Nutrition as Mineral Balance

This understanding brings us to repeat many of the concepts already presented. Soils out of balance will attract weeds to help bring them back into balance. Plants out of balance will attract insects to devour them, thereby eliminating them from the animal food chain. How I remember the joy and pride an Amish farmer displayed in an educational seminar as he showed corn taken from his field with no insect damage and alongside he displayed common weeds from the same field riddled with insect damage. He had succeeded in balancing his field to the point where the theory was working.

Why does it work this way? This theoretical view is based upon the premise that nature, God, or whatever name you wish to use, has an innate wisdom behind it. It views creation as an extension of purpose and order which, if left to its own designs, can remedy itself.

That's why we say, if you will stop using the products which are producing damaging effects on crops and cropland, you will soon begin to see Mother Nature recover the damage. Depending upon the damage, this recovery will progress more or less quickly. Man has discovered many of Nature's laws and has chosen to recognize some more than others. Whether due to greed, ego, or just plain stubbornness, ignoring these laws will only lead to failure or, at most, incomplete success.

Dr. Reams, in addition to his findings in agriculture, made significant findings in the area of human nutrition. Through the process of measuring a person's urine and saliva for sugars (refractometer), salts (using a conductivity meter), and nitrogens (ammonias and ureas), he was able to pinpoint the nutritional deficiencies causing the particular problems complained about. It was through searching out solutions to these complaints that he discovered the importance of food mineralization.

Dr. Reams discovered that one mineral the heart needs in abundance is arsenic, preferably arsenic of phosphate. He found that calcium combines to form over 240,000 different compounds in the human body and the lungs need a wide assortment of minerals, more than any other organ in the body.

To supply the heart with arsenic, Dr. Reams would recommend a person eat asparagus. Naturally high in arsenic, eating asparagus two or three times a week would help heal a damaged heart. Eating it too often might kill a person because too much arsenic can become counterproductive. Additionally, the arsenic found in properly fertilized asparagus differed greatly from that found in improperly — usually conventionally — fertilized asparagus.

These findings are not new. Rutgers University published findings comparing the nutrient (mineral) content of foods grown on one type of soil compared with those grown on other types. The specific mineral content of foods varied anywhere from 1 to 18 times the norm depending on the mineral content of the soil. This is the difference identified by Dr. Reams and related to soil and plant fertility.

The organic movement received a great promotional boost from the public's fears over the use of the chemical Alar, used on apples. Whether or not these fears were overblown can be debated. Chemical farming adherents may argue that the data presented was not supported by animal studies. They may be correct. Chemical farming opponents will argue that the human system isn't capable of detecting the long- or short-term problems or conditions caused by the use of this chemical. Government agencies, including the Environmental Protection Agency, have not evaluated a fraction of the chemicals sold on the market today. Additionally, they haven't addressed the question, "What happens when chemicals combine with other chemicals?" Dr. Philip Callahan, in association with Louisiana State University, tested most, if not all, of the chemicals then on the market. He concluded that all were harmful to the soil. Could this be why his findings have been ignored by the entomology community and only embraced by the Department of Defense?

Society may come closer to knowing the actual response such chemicals have on soil, plant and animal life when new equipment is developed and utilized. Any product used on the crop or soil which lowers the vitality of either should be viewed as suspect. With the technology available to increase the vitality of soil and crops while at the same time lowering the vitality of weeds and insects, there is little reason to look to toxic chemicals for "rescue." Rare cases of plagues of locust or grasshoppers may still require some type of disease, predator or poison to stop the damage, yet perhaps even these have their purpose.

In an apparent confirmation of Dr. Reams' premise regarding disease and other conditions relative to nutrients deficiencies, Dr. Arden Andersen, author of *The Anatomy of Life and Energy in Agriculture*, has several pages showing different crops, the problems they encounter, and the mineral deficiency behind each of them. The following is reproduced, with his permission, from his book:

Insect, Disease or Condition in Crop	Vector	Sequential Nutrient or Disease Deficiencies
ALFALFA		
Common Leaf Spot	Fungus	P, Ca, (spots from potash excess, yellow due to either magnesium excess or nitrogen deficiency which excess magnesium causes)
Anthracnose	Fungus	P, Ca, vitamin C
Bacterial Wilt	Bacteria	P, Ca, vitamin C, Fe/Cu, Se/Co, vitamin A
Downy Mildew	Fungus	P, Ca, vitamin C, Cu, Fe
Fusarium Wilt	Fungus	Ca, Cu, P, Fe, Mo
Rust	Fungus	P, Ca, vitamin A, Co
CELERY		
Aphids	Insect	Ca, P, Fe, Cu
Black Heart	Rapid Growth	Ca, P, B, Co
Cabbage Looper	Insect	P, Ca, vitamin C, Fe
Common Celery Mosaic	Virus	Ca, P, vitamin C, Co, Se
Damping Off	Fungus	P, Ca, vitamin C, Mo
Early blight	Fungus	P, vitamin C
Late blight	Fungus	P, vitamin C
Variegated Cutworm	Insect	Ca, P, Cu, Mn, vitamin C, vitamin E
Fusarium Yellows	Fungus	Ca, P, Cu, Fe

STRAWBERRIES

Aphids . . . . Insect . . . . Ca, P, Fe/Cu

Black Root Rot . . . . Unknown . . . . Ca, P, vitamin C

Flea Beetle . . . . Insect . . . . Ca, P, Fe/Cu

Leaf Spot . . . . Fungus . . . . Ca, P, Se

Nematodes . . . . Insect . . . . Ca, P

Powdery Mildew . . . . Fungus . . . . Ca, P, vitamin C, Co

Stem End Rot . . . . Fungus . . . . Ca, P, vitamin C, Mn

Spittlebugs . . . . Insect . . . . Ca, P, vitamin C, Fe

Leafhopper . . . . Insect . . . . Ca, P, vitamin C, Co, Se

Weevil/Clipper . . . . Insect . . . . Ca, P, Fe, Cu

Verticillium Wilt . . . . Fungus . . . . P, Cu, Mn

Viruses . . . . Virus . . . . Ca, P, vitamin C, Co,
Se, vitamin E

White Grubs . . . . Insects . . . . P, Ca, vitamin C,
Mn, Co,Cu

POTATOES

Black Scurf . . . . Fungus . . . . Carbohydrate

Colorado Potato Beetle . . . . . . Insect . . . . Ca, P, vitamin C, Cu, Mn

Common Scab . . . . Bacteria . . . . P

Dry Rot . . . . Fungus . . . . Ca, Co

Potato Flea Beetle . . . . Insect . . . . vitamin C, Ca, P, Fe,
vitamin E, Cu, Mn

Potato Leaf Hopper . . . . Insect . . . . Ca, P, Mn, Cu, Fe

Potato Leaf Roll . . . . Virus . . . . P, Ca, Cu

Soft Rot . . . . Bacteria . . . . Ca, K, B

Variegated Cutworm . . . . Insect . . . . Ca, P, Cu, Mn,
vitamin C, vitamin E

White Grub . . . . Insect . . . . P, Ca, vitamin C,
Mn, Co, Cu

Wilt . . . . Fungus . . . . P, Cu

Chapter 17
Water

What's to write about water? Simply sink a well and pump up all the water you need. After all, pure water is a part of the rural landscape and the right of every American.

Until a few years ago, this was the case. Rain washed through the layers of purifying sand and soil settling into aquifers ready for use. Our only problems with the water consisted of having to purchase a water softener to get rid of the hardness or deal with the saltiness, iron or sulfur.

Background

By the mid-1980s, the EPA had reported that thousands of wells had been closed because of contaminated water. Nitrates and more than sixty farm pesticides had been identified.

The problems were traced to many sources including petroleum wastes, strip mine operations, land fills, chemical dumps, spills, and eventually to even the relatively small amounts used as yearly applications in farm operations. In many cases, people suspected problems before testing was initiated. Medical problems showed up in areas statistically above the levels expected. However, just as in the case for tobacco and cancer, cause and effect could not be positively proven. Profits had been made and people were suffering, but the medical links were elusive.

For years, little monitoring was done, and landfills became the depository for just about any and all chemicals. Current laws prevent using landfills for much beyond household garbage and trash. Toxic waste landfills now require extensive foundational construction of clay and other impervious materials so as to prevent the leaching problems which contaminate ground water. Because of the costs involved, including combatting the legal maneuvers by opposition groups, other approaches to disposal are being utilized including recycling and incineration.

Groundwater Contamination

How bad is the problem? Considerable and growing. Growing not necessarily because more contamination is occurring, but because we are now testing for contamination. *The Philadelphia Inquirer*, for instance, has reported that more than 40 small water systems in Pennsylvania are polluted and don't meet safe drinking water standards. It is fairly common to read about farmers and even small communities forced to purchase "outside" water due to contaminated wells.

Indiana, Michigan, Ohio, Illinois, Wisconsin, Minnesota, Iowa, Nebraska, Pennsylvania, Delaware, Washington and California currently have the most contamination of nitrates and pesticides. Texas, Oklahoma and Kansas have considerable nitrate contamination while the coastal states, including Louisiana, have heavy amounts of pesticide contamination. In this case, the term pesticide includes all farm-use toxic chemicals, such as pesticides, herbicides, nematicides and fungicides. The threat is real and not merely imagined.

More than 20,000 hazardous waste sites have been identified (without counting the thousands of illegal dump sites), 90,000 landfills, over 10,000,000 underground storage tanks, 200,000 storage lagoons, millions of home septic tanks, hundreds of thousands of wells — gas, oil, water, etc. (a source for infiltration to ground water reservoirs), and thousands of farm manure holding lagoons all contribute to the groundwater contamination problem. Household wastes dumped down the drain can be as great a concern as any other contamination source. Few household owners realize that the septic system works only because of the microbial life contained in the sewage sludge. By decomposing the sludge, the microbes are able to change the raw sewage into a decomposed, or composted form which can then be readily used by soil or plant life. This process is similar to that used on the farm to break down manure.

System Problems

Problems occur, however, when something is added to the system which interferes with the breakdown process. In the case of household septic systems, the breakdown can occur from the dumping of motor oil, household pesticides, paints or paint thinners, flammable liquids, chlorine bleach, certain cleaning supplies, or other contaminates which destroy the microbial activity in the septic tank, in turn destroying the breakdown action. When this occurs, the septic tank must be pumped and the contents hauled away for disposal elsewhere. Once contaminated, septic systems have difficulty recovering to their intended level of biological digestibility. It is most important that communities and households educate themselves on which products are capable of contaminating septic tanks and eliminate these from regular use or learn how to dispose of them properly.

Similar problems occur on the farm. Much of our livestock intended for human consumption is raised on medicated feeds. The low-level antibiotics are required because the animals are basically unhealthy (sick) due to the conventionally used fertility approach. Lacking the necessary vitamins and minerals needed for proper health — plus the added animal stress brought on from consuming molded or otherwise tainted and low brix feed and living in unnatural confinement with limited sunlight and oxygen — the animals simply get sick. In order to raise the animals to full term, medicated feeds are prescribed. Now, the animal can be sold at a profit, and everybody is happy.

This approach becomes a nightmare when a consumer contracts some type of disease and no longer responds to medication prescribed by the family doctor due to the resistance factor obtained from eating meats grown in this manner. Recently, *E. coli* contamination of unpasteurized apple juice resulting in death to a consumer cries out to the entire agricultural/food/health system. How did *E. coli*, a large intestine bacteria, get in cider, and why is it so virulent?

Manure Breakdown

Manure produced in a typical medicated feed and confinement operation does not contain the proper bacteria. Just like the above discussion on household sewage systems, manure will not decompose properly when the necessary bacteria are missing or when the sugars normally found in manures from animals fed healthy, high-sugar crops are low. These two variables hold a key to the widespread manure problem faced by most conventional livestock farmers today.

Most manure produced today is in a putrefied state and not easily converted or used by soil or plant. When this material seeps into groundwater, wells are also contaminated. In some cases, contaminants will enter the ground water by following the same entrance used to bring water to the surface — the well itself. Because of this, the EPA has established a Wellhead Protection Program to encourage the protection of wells. Included as part of the Safe Drinking Water Act Amendment of 1986, it was designed to help state and local zoning boards in making decisions to protect ground water by protecting the surface and subsurface areas around a well.

Correcting tainted ground water is not a simple or inexpensive matter. The USDA estimated the costs to test water for nitrates, 2,4-D, atrazine, Temik and Lasso at an average of just under $200 per test. Prevention includes removing underground tanks, properly disposing of chemicals in approved landfills, and capping unused wells. You would be well advised to take these steps as quickly as possibly because if contamination is discovered, the liability costs have just begun.

Groundwater Cleanup

How costly is it to clean up the groundwater? In 1977, a storage tank in a northeastern state leaked almost 3,000 gallons of gasoline before being detected. Soon after, residents began complaining of water which smelled bad. Later, the town was forced to close the well. The cleanup consisted of local, state, and federal cooperation and was completed in 1985 at a cost of over $3 million. That's $1,000 per gallon of gasoline. This is only one of thousands of wells contaminated by a host of contaminants.

According to Superfund (Comprehensive Environmental Response, Compensation and Liability Act) guidelines, any "potentially responsible party" must pay for cleanup costs. They consider a potentially responsible party as being anyone who currently owns or operates the land plus any past owner or operator of the land since the contaminant was disposed. In this regard, the EPA is not interested in proving who was negligible, only who owned or managed the land. If you are a landowner, you are responsible for the actions of your renters. Even lenders are beginning to bear the financial responsibility.

Groundwater contamination is a serious issue not only because of the liability and costs of cleanup, but also because of the health toll it takes on those consuming the tainted water. As stated earlier, thousands of individuals have suffered medical problems due to drinking tainted ground water. Overlooked, however, is the effect on countless livestock who also drink the water. Livestock suffer the same decreased performance syndrome as do people, except they can't complain. Their performance goes down with no identifiable cause. Conventional analysis measures the water for nitrates or coliform bacteria, but not for atrazine or other poisoning. Much production is lost with nothing to account for it.

Farm Treatment Options

What can be done to correct contaminated groundwater? Besides preventing the contamination itself, the farmer has relatively few options. Basically, they fall into several water treatment categories: activated carbon filtration, ion exchange, reverse osmosis, distillation, ozone or hydrogen peroxide, chlorination, or a combination of these.

Granulated activated carbon has been the longtime standard for removing pesticides and other contaminants. It holds these in its structure. These filters can also have negative side effects. Activated carbon is an excellent environment for bacterial contamination. In addition, it not only can grab and hold on to contaminants, but it can also release those contaminants unexpectedly. This can happen when the water changes temperature or pH rather quickly. Drinking water from a filter which is releasing its hold means you are drinking water many times worse than the untreated water.

Farmers can take care of this problem relatively easily by periodically back flushing their carbon filter. One top rated manufacturer of sink-side units utilizes a solid carbon filter which effectively filters throughout the life of the filter — 1,200 gallons.

Ion exchange units function similarly to water softeners. To eliminate hard water, you add salt (sodium chloride) to the softener. The softener uses the sodium ion to exchange for the calcium ion, and the water is no longer hard. Other ion exchange mediums exist to further this process. Usually using the ion exchange approach consists of renting the ion beds from a commercial source which periodically replaces the beds keeping your water supply always at the level you want it. The reverse osmosis and distillation methods work well, especially at low levels of output. The reverse osmosis unit uses about seven times the water that is turned out just to keep the membranes clean. The distiller, of course, uses electricity to operate.

Ozonization units inject ozone directly into the water line. Some are designed to remove the ozone later, before consumption by passing it through a carbon filter. Studies have shown that by coupling an ozonization unit with an ultraviolet bulb in the water line a tremendous oxidizing environment capable of destroying pesticides is created. The EPA has given this technology its "best available technology" rating.

Walter Grotz has told countless stories of people and livestock operations which have been benefitted from treating water with hydrogen peroxide. Medicator-type units inject food-grade 35 percent hydrogen peroxide directly into the water line. Both ozone and peroxide will kill bacteria on contact. Walter Grotz believes the peroxide adds beneficial oxygen to the water. There is a group of medical doctors which is experimenting with injections of hydrogen peroxide directly into a person's veins. This treatment approach also requires attention to consuming antioxidants such as vitamin E and vitamin C. This attention might also be recommended if a person was using a peroxide water treatment system.

Conventional chlorination units work by installing a holding tank in which water stands in contact with chlorine added to the water. After a period of time, the chlorine penetrates the outer membrane of the bacteria and kills it.

Many farmers would want to consider a combination approach in treating their water, especially if it was also being used for home use. For example, they might use a medication injector to add hydrogen peroxide to the water plus a carbon filter to pick up pesticides. If the carbon filter were placed after the peroxide, the peroxide would eliminate any bacteria from growing in the filter.

Water is crucial to every living thing. Consumption of good water is important to the healthy functioning of the body at the cellular level. It is imperative that water be cleaned prior to drinking by you and your livestock.

Chapter 18
Forage

Forage Quality

The conventional approaches to fertilizing forages have led to too much imbalanced, poor-quality feed going into our livestock. Under the current economic conditions, this has prompted everyone to become concerned over the cost of raising livestock, especially the cost of supplements required to make up for the deficiencies in the forage.

When farmers inquire about methods of raising better (more nutritious) alfalfa, the conventional answer comes back with recommending 0-0-60, keep the pH up, cut by the blossom, herbicide the weeds, use 18 pounds of seed per acre, and all the other wrong or *wrongly reasoned* advice.

The failure of standard forage fertility programs is appalling. The authors have received frantic calls about alfalfa weevil infestations that won't go away. We suggest molasses and 35 percent food-grade hydrogen peroxide to those who have emergencies. Sometimes it works, sometimes it doesn't. The problem with this program is finding either product. The rates to use are also variable.

Sometimes weevils are left in the field, but cease doing any more damage. This means they are either hampered by the mixture, the brix have risen too high, or both. Some literature suggests as much as a pint of 35 percent hydrogen peroxide per acre, yet we have seen other reports indicating as little as 1 or 2 ounces an acre. Usually, the 1- or 2-ounce level has been accompanied by molasses, sucrose, and/or liquid calcium at 1 to 2 gallons per acre. Although this may be a good alternative way to make the weevil want to leave, we can't lose sight of the fact that the weevil's presence is a sign of low brix in the crop. So regardless of your reaction to the weevil, spraying, early cutting, or whatever, you have to come back with nutritional sprays or the problem may remain for the next cutting.

Recent issues of farm magazines have contained numerous articles about forage quality. These articles describe the various tests such as NDF, ADF and crude protein used to measure forage quality. They also talk about a new RFV (relative feed value). Many field reports indicate biological growers producing forages with very high RFV's. They also report high brix and no insect attack. We believe a concentration on calcium and phosphorus availability, supplemented with sulfur and trace minerals, is allowing the plant to produce sugars of many different kinds. High brix readings are correlated with high sugar readings, but what about the high RFV reading which depends on ADF or NDF fiber reading?

The ADF and NDF readings are supposed to be kept low while keeping the protein up. It seems that you want lots of cellulose and hemicellulose fiber which can be digested by rumen bacteria, but not too much lignin fiber which is indigestible. Textbooks verify that cellulose and hemicellulose are formed from the same class of carbohydrates and sugars as our popular soil "glue," namely polysaccharides. These polysaccharides are long chains of six-carbon ring structures of a specific glucose which make up fibers of the plant's cell wall. By contrast, lignin is formed from alcohol-type substances rather than sugars. Our conclusion is that our methods described above allow or force the plant to produce more cellulose and hemicellulose because of the large amounts of sugars available. This is why we downplay crude protein and emphasize brix (sugars or carbohydrates). This way you get both factors which give high RFV, digestible fiber and energy.

Even if you have great forages in the field, you can still lose quality before it reaches your livestock. The first way is to chop too short. Not only do you lose more fluids (sugar and energy), but you can lose your butterfat potential. The butterfat level is largely a function of the tickle factor of longer stems in the rumen plus the natural bicarb production of the cow. Try to stay at an inch or longer.

Another way to hurt your crop is to not ensile it fast enough. The faster you get it in the silo, pit or bunker and get the oxygen out of it, the quicker it will ensile and retain nutrients. Although high-brix forages should ensile without inoculants, it's a matter of chance because of reduced natural bacteria occurring on field crops. Consider an inoculant. A third way is to fail to check your refractometer before cutting.

Several factors could temporarily drop the brix reading. Plant stress due to moisture, temperature or low pressure could trigger the plant to move the sugars to the roots. Early morning or late evening cutting may also reduce the brix. Look for the brix to be best from 12 p.m. to 4 p.m., but always check it yourself.

More permanent damage can be done by playing the high brix game on a late fall cutting. If you are new to a biological program, your crop may not be able to sustain adequate sugars in the leaf and the roots for winter survival. Be cautious about going for maximum brix on that late cutting because it won't allow time for regrowth and remanufacturing of adequate sugars. (You must have adequate calcium and phosphorus to accomplish sugar production.)

It is just as important to ground, crown, or foliar feed a forage crop as it is a row crop. Typical liquids include molasses, liquid calcium, fish, seaweed, hot mixes or dry-soluble fertilizers in solution. Feeding can be done during growth, between cuttings, or a creative farmer can spray the crown's while cutting. Use your imagination and monitor the brix.

Chapter 19
Livestock Nutrition

College courses are taught to convey the information needed to properly raise and care for livestock. However, the basic feed knowledge seems to have been somehow perverted into a system of requiring hormones, antibiotics, excessive proteins, and excessive silage. This perversion has resulted in a variety of problems involving early burnout of dairy cows, acidosis, anemic-looking eggs, chicken with no flavor, rejection of U.S. meats by foreign buyers, and anti-cholesterol (egg and dairy product) publicity. It's time livestock producers start looking at changing their production practices which may help eliminate these complaints.

Water

The first step is to clean up and balance the water going into the livestock as covered in Chapter 17. This can be accomplished by a variety of methods involving charcoal filters, injectors for chlorine, hydrogen peroxide, vitamin C, subtle energy devices, iron filters, and even magnets. Whatever your choice, it should be done as soon as possible to allow the animal to take maximum advantage of whatever changes you make. Crop improvement takes time, but improved water can achieve great results almost immediately. You can try hooking up a meter to monitor live-stock consumption before you make any changes. Then check consumption after you improve the water. The proof is in the volume of drinking.

Feed

The next step to livestock nutrition is to clean up the feed. This may initially be nothing more than the use of gentian violet or GV-11 (an anti-fungal product) to minimize the damage being done to livestock liver and immune systems from mold and afla-toxins. GV-11 has been approved for use in chicken feed only. Why hasn't it been approved for other animals since moldy feeds are a common problem?

The best method is to start changing field practices and products so that you can grow more nutritious food with little or no contamination from mold, herbicides and pesticides. The major emphasis is putting calcium, phosphorus, carbohydrates, true protein, trace minerals and vitamins back into the feed by the use of scientific biological agriculture.

Because of the poor quality of contaminated feed and water, most animal nutritionists have resorted to pushing protein (nitro-gen) as the answer to production. The results are the same as using excess nitrogen in the soil. You get results by increasing the electrolyte in the system, but you burn out the system without correcting the initial problems. Although some results can be achieved by reducing protein in the ration and adding fiber and energy, the final answer is raising the carbohydrates and true, digestible proteins in the feed.

One of the major problems involves the basic calcium and phosphorus minerals. Modern high-production animals usually need supplementation. This opens the door for marketing abuses. One problem area is the use of "fertilizer-grade" versus "feed-grade" phosphorus in the marketed supplements. As with fertiliz-ers, the better the quality and energy level of the mineral being fed, the better the end result. Another is the presence of signifi-cant amounts of limestone (calcium) in soybean meal which imbalances rations. The net effect on many farms is for livestock to always be deficient in phosphorus due to excess calcium.

When the farmer tries adding or free-choicing phosphorus, the amount needed or consumed seems too costly. The trick is to lower the calcium being added to the ration, to balance it down in comparison to the phosphorus. Of course, there are rations in all stages of imbalance, and no one should remove calcium by guessing. One way to bring the calcium-phosphate balance into line prior to feeding minerals is to discontinue the practice of sowing straight alfalfa. You would be better off if you sowed an alfalfa-grass mixture. The alfalfa contributes the higher calcium levels while the grasses contribute the higher phosphorus levels.

Free-choicing of minerals and additives is a controversial concept. It is said to be impossible by some experts for an animal to choose, free-choice, what it needs. Why then, do animals chew bark, eat dirt, drink from corral urine pools, and crib? Why do they stop that activity when given specific supplements? There is a strong case for free-choicing, and each producer should explore it for himself.

Also, be aware that nutritionists will prescribe large doses of certain trace minerals to counteract high nitrates or other contaminants in water or feed. If you clean up the contaminants, be sure to tell your nutritionist so he can program accordingly. Otherwise, you may have a toxic reaction to an excess of a specific trace mineral in the near future.

Nutritionists often rely heavily on an approach which involves the use of buffers (sodium bicarbonate) to raise the pH of the rumen or digestive juices. This practice is an admission of improper functioning of the animal usually due to lack of long-stem fiber stimulation in the rumen. This lack is due to feeding short-chopped, soft haylage instead of long-stem, scratchy hay. All dairy feed programs should contain some dry, long-stem mixed hay (grass and alfalfa).

Salt

Salt and trace minerals are easily mishandled. Never force the animal to choose between salt or trace minerals by offering them both in one form. Always have free-choice, loose, white salt available besides whatever else is offered in the way of salt/trace mineral mixtures. Most dairy or beef cows cannot get enough salt from a block during hot weather.

One of the best dairy setups would be ground ear corn coming out of a wooden crib, oats from a wooden bin (treated with diatomaceous earth), long-stem alfalfa/grass hay, and some corn/hay silage out of a non-metal silo. The grain mixture could contain whatever other items necessary to balance the ration such as salt, calcium, phosphorus, kelp, diatomaceous earth, trace minerals, etc. Very little protein should be required. Always seek out local feedstuffs, such as brewer's grain, vegetable or fruit pulp, and linseed meal.

Housing

Metal storage buildings, containers or silos used for storing feeds are questionable because of the electrical effects between an organic compound (feed) and the metal building. Better suggestions are the use of wooden cribs or bins and cement silos or bunkers. Placement is also important as earth magnetic energy lines can affect feed storage.

Location and type of building can also be significant. Both standard stray voltage as well as earth magnetic phenomena can have serious effects on livestock. Recent work by electronic scanner operators has shown the potential for metal buildings to draw in other electromagnetic energies, such as radar or microwaves. When all else seems to be right, but the animals aren't, consider the building. In cases where stray voltage or other electromagnetic interference is a concern, the metal building should be grounded. Ground each panel including the roof. If a portable radio plays better with the antenna touching the side of a metal building rather than in the air, you have a grounding need.

Lighting is also an often unnoticed variable which plays a crucial role in animal health. In grandpa's day, he milked the cows and turned them out to pasture. In modern days, we milk the cows and return them to the barn until the next milking. Animals in confinement seldom are exposed to natural sunlight.

John Ott, a researcher, has written several books on the concept of health and light. His findings indicate a direct correlation between malnutrition and the light source. He has found that the use of full-spectrum light fixtures can produce dramatic effects on livestock as well as humans. Several companies have worked with his patents.

Special Tips

The following is a list of special tips for livestock:

Poultry

1. For better water — carbon filter and ozone or peroxide.
2. For low-cholesterol eggs — full-spectrum lighting and diet.
3. For good health — no meat scraps and good ventilation.
4. For shell strength — diatomaceous earth and/or apple cider vinegar to release calcium.
5. For fly control — diatomaceous earth
6. For proper egg yolk color — free-ranging, alfalfa meal, or kelp in ration.

Swine

1. For better water — carbon filter and ozone or peroxide.
2. For rhinitis — zinc and ventilation.
3. For good health, digestion and manure control - biologically grown corn, soybeans, vegetables, alfalfa, and good-quality feed supplements.

Dairy

1. For better water — carbon filter and ozone or peroxide.
2. For digestion — B vitamins, cobalt, and other quality supplements.
3. For worming, mineralization and buffering — diatomaceous earth and/or clays such as Dyna-min.
4. Avoid iodine in teat dips (use peroxide instead).
5. Use probiotics instead of antibiotics.

6. Use vinegar and/or molasses instead of urea or anhydrous o n silages, poor hay, etc.
7. Use well-cared for and fertilized pastures whenever possible.
8. Use the Voisin method of intensive pasture management.
9. Use corn, soybean or other oils in the ration for dry, dull coats.
10. Feed your calves, heifers, and dry cows better (correctly).

Thanks go to Jerry Brunetti of Agri-Dynamics for the following information regarding animal nutrition. Although this could easily be construed as a product endorsement, the information explaining the product's mode of action and the need for corrective action in most livestock is too valuable not to pass along. Other clay products might perform similar functions.

Dyna-min is a mined clay product rich in minerals which benefits dairy cows, swine, and to a lesser extent beef. The following will explain the components of why this product enhances feed efficiency and herd health and explain animal nutrition — and its link to soil fertility — in the process.

Calcium Availability

NASA tested Dyna-min as part of an investigation into what could be done for astronauts in orbit. Once in orbit they lose bone mass. NASA concluded that there is more bone development when using Dyna-min than from calcium limestone. The reason is the way the clay crystal is structured. It was formed with tremendous heat and calcium is loosely attached to the center. The product has a calcium level of about 4-5 percent in content.

Buffering Capacity

Dyna-min has acid-alkaline buffering capabilities. High producers generally use too much sodium bicarbonate in feeding steers and dairy cows and the result is the animals don't produce enough bicarbonate from their own saliva, which is the result of cud chewing by having adequate "effective fiber." Dyna-min will run a pH of about 7 and it produces a buffer balance in the gut.

If you took this product and compared it to sodium bicarbonate in a chemical test, sodium bicarbonate would react more quickly as a chemically ionic substance with hydrochloric acid. Sodium bicarbonate will fizz because it has a pH of 10. Dyna-min soothes like Pepto Bismol through the entire digestive track and produces a nice acid/alkaline balance. If you let animals free-choice as a part of their feeding, they will know how much to take to balance the pH in the gut. It performs as both an absorbent/adsorbent of excess acidity.

Detoxifying Agent

Another factor is the cation exchange capacity whereby a strongly charged negative clay matrix with loosely attached cations exhibits a detoxifying property. These cations in the clay are easily exchanged for positively charged cations which tend to be your toxic substances, like ammonia, which are captured and carried out of the system.

There was a test done on well water which had 42 parts per million nitrate nitrogen (the limit tends to be 10 or below to be safe). They took 1 liter of this water, put in 1 ounce of Dyna-min and put it on a magnetic stirrer for 24 hours. After 24 hours it eliminated 100 percent of the nitrate from the water. The material has an incredible scavenging capability for free nitrogen, be it nitrate nitrogen, or nitrogen, or ammonia nitrogen.

Nitrogen is a big problem for cows (particularly dairy), because what you create is a condition known as BUN, blood urea nitrogen. Elevated BUN conditions are not natural; they come about from the feeding process. The cows won't breed, they abort, develop mastitis that won't clear up, and experience poor feed conversion. Dyna-min also has detoxifying properties that absorb mycotoxins similar to some synthetically produced calcium and sodium aluminum silicate products. Mycotoxins are poisonous excretions of certain molds that are very destructive to the liver, kidney, bone marrow, respiratory tract, reproductive organs, and the immune system. They are predominantly found growing on starches and are the result of unhealthy farming practices including excess nitrogen, herbicides, fungicides, poor harvesting techniques and faulty storage systems.

Enzyme Production

Enzyme production, or feed synthesis, is very important for productivity and health in livestock. A test was done by Biotic Research Labs. They fed livestock Dyna-min for a period of 90 days and took manure samples every week, analyzing them for amylase which digests starch, lipase which digests fat, and protease which digests protein. These three enzymes were analyzed from horses, cattle and poultry. At the end of 90 days, without exception, the enzyme levels came up and normalized in all three species. They actually normalized much sooner than 90 days, but the experiment was continued for 90 days to see how well the levels held. The trial consisted of four quarters, each 90 days in length, lasting one year. So, the trial consisted of 90 days with Dyna-min, 90 days without, 90 days back on, and 90 days without. Every time Dyna-min was removed for the 90-day quarter, the enzyme levels dropped and the parasite eggs returned to their original starting levels.

Trace minerals are a major part of an enzyme's characteristic. There are no enzymes that do not have a trace mineral which is a part of the activity of that enzyme. Enzyme production is very important. Animals get most of their enzymes from the food they eat. Nature puts enzymes in feed so animals will be able to digest those nutrients within the feed itself. Of course, these enzymes are only viable in uncooked foods, but then animals have not evolved yet to the point where they cook their foods. We do it for them.

Digestive enzymes are secreted out of the pancreas. In people the pancreas is particularly challenged, because the body is producing all of the enzymes necessary to digest the food instead of getting them out of the food that's consumed. There is what is known as an enzyme reserve. The body has only so much capacity to manufacture enzymes. By eating mostly cooked foods we use up this enzyme reserve faster than when we eat raw foods.

Enzyme Importance

We challenge animals by feeding them food that's not indigenous to their diet. For example, we feed grain to cattle (dairy and steers) and swine, but these aren't what their native relatives are used to eating in the wild (forages, forbs, roots and mast). Another way we challenge animals is in the way we grow the feed. Current fertility practices ensure that crops are grown such that there is very little enzyme activity in the feed. Additionally, we put it in containers, such as silos, which are often maintained at only 40 percent moisture and it ends up being mostly sawdust, again without any enzyme activity. Animals don't have the proper digestive efficiency on that feed because the whole package isn't there anymore.

A statement about feeding grain, ground seed, might be in order. Once a seed matures, it contains a package of enzyme inhibitors which prevent digestive enzymes from being available. That's one reason why we roast grain for hogs; otherwise you'd get very poor feed conversion. These naturally occurring enzyme inhibitors also keep the seed from sprouting prematurely when the right sprouting conditions are not present. What happens when you put that seed in a moist, warm environment? It takes in water, the enzyme inhibitors and phytates are destroyed, the seed germinates, and the sprout comes out. When animals (including ourselves) eat grain (corn, barley, etc.) they're eating the enzyme inhibitors contained in the seed which not only inhibit the beneficial digestive enzymes, but which also contain substances that will tie up a lot of the nutrients, such as phytates.

There is only one animal that should be eating whole grain — poultry. Why can a chicken eat whole grains and not a hog or cow? The chicken's crop is the sprouting container. The seed goes into the moist crop and swells up after which it goes into the gizzard. The moisture that goes in renders the enzyme inhibitors out. Now you have a seed that has all the food enzymes without the food enzyme inhibitors. You could also get rid of enzyme inhibitors by cooking them out. But when you cook the inhibitors out, you also destroy the digestive enzymes.

The best way to deal with food production on a farm is to soak the grain. There are farmers in Pennsylvania who soak wheat and rye before giving it to their dairy cows. You would not believe the difference in milk production. This is because the grain has the digestive enzymes needed to digest itself and because it's gotten rid of the enzyme inhibitors. Dyna-min will help synthesize more enzymes in the animal. It also increases enzyme output of an animal and the response is remarkable.

Enzyme Research

To show how important enzymes are, the Price-Pottinger Nutrition Foundation conducted nutrition research with cats in the 1940s. One of the experiments showed that what went into an animal and then went on the ground after going through the animal is contingent on the enzymes being in the food that the animal consumes. They studied multiple groups of cats; one group received raw milk from the cow and the other groups received pasteurized milk, condensed milk, and evaporated milk (all cooked milk). They then collected the manure from each group of cats and put it on separate sections of ground, dug it in, and then planted bush string beans on the plots. The beans grew twice as high and put out twice as many beans from the manure from raw milk versus the manure from cooked and processed milk. This finding shows that the enzymes in the raw milk ended up affecting the quality and the health of the green beans via the quality of the manure.

When you're spreading manure on a farm, the quality of that manure has a lot to do with the health of the crops and the yields that you can get from the crops. How manure is handled is one thing, but how the manure is made is probably just as important if not more important. Most people complain about the putrid smell from manure. The reason why you have a putrid smell is because you don't have the right levels of enzyme activity in the gut of that animal that made the manure. Manure shouldn't be offensive, perhaps strong if you're near a lot of it, but it should not have a putrid, awful smell to it. Of course, a lot of that has to do with the way its stored. Anaerobic pits will cause the putrid smell problem, because anaerobic pits produce indoles, skatoles, mercaptans, sulfides and ammonia. These are substances not found in an aerobic soil and are toxic to beneficial soil organisms.

What about antibiotics? They can kill the gut organisms that help produce the enzymes. Microorganisms require nutrients because they are the *livestock* of the soil and the *livestock* of the gut. In some ways you do not feed the animal, you do not feed the plant, but you *do* feed the digestive system of the respective plant and animal. The way you do that is you feed the livestock of those respective kingdoms — the microorganisms.

One of the reasons you get a reaction from biological fertility programs is because you're feeding microorganisms that are able to mobilize the macronutrients which may be present to supply the calcium, magnesium, phosphorus and potassium, as well as the numerous trace elements, but which were not previously available to the crop. It is important to look at the enzyme factors in terms of getting microorganisms to work on your farm. Getting the manures to increase enzyme transfer up from the soil to the crop and the crop to transfer food enzymes to the animals, and the manure to transfer back to the soil is the way to keep the cycle going. Of course, this is the objective of compost and compost tea — to provide the microfloras that perform as catalysts in the rhizosphere of the plant.

The Pottinger study is fascinating. One experiment they performed in the 1940s showed that the animals' digestive systems knew exactly when and what food passed through their gastrointestinal track. They found that there are natural monitoring systems in the gut which pick up information as to exactly how much protein, fat and starch is coming into the alimentary canal and how much of that food contains the lipase, amylase and protease to digest the starch, fat and protein. The pancreatic output will make up the difference. The "system" knows.

This research made me realize why soaking grains for dairy cows works and why it works so well. They turn grain into cereal grasses in about five days. What they are doing in essence is doubling the dry matter of that grain in that same five days, but they are also increasing all the other factors of that grain such as chlorophyll content, enzyme activity, vitamin levels and available micronutrients. The results are being seen in Europe. As an example, there are some very exciting experiments with animals that are recovering from surgery. The recovery time after surgery for an animal is only a fraction of the normal recovery time on large animals like horses that are being fed these cereal grasses. Again, that's not surprising because animals in the wild eat these things normally anyway.

The normal amount of time that the grains need to sprout is 48 to 72 hours, depending on the grain. You should soak the grain and water should be added every 24 hours, at a minimum, in a container that drains. Add 1 ppm of iodine to the water to prevent mold growth. At 48 hours there should be the beginning of a sprout breaking through the kernel, but at 24 hours you will get a swelling and oftentimes you will get a removal of the inhibitors. So 48 to 72 hours will generally do it. They normally soak the seeds 24 hours before they put them on trays and then expose them to a full-spectrum light or sunlight. It's all automated and the humidity and temperature are controlled in the commercially available sprouting systems.

Where would you rank fermented (but not moldy) high-moisture corn since it goes through some sprouting and fermentation? In my opinion, it's better than regular dried, shelled corn, but it's still a lot of starch, and cattle are not indigenous consumers of carbohydrates in a starch form. So what happens when they eat more than a normal amount (a normal amount is next to nothing)? (The only carbohydrates cattle would find in the wild are the seeds that they would pick off the heads of forages.) You're going to get a lot of lactic acid production. It's going to create an inhibitor-like effect in the rumen, killing off bacteria that break down the real source of energy for the cow which is fiber, cellulose and hemicellulose, pectins, sugars and beta-glucans. We do not have any herds that are on a full-forage diet, because milk production is a matter of getting enough energy into the cow. If you're breeding a cow to produce 25,000 pounds of milk, where are you going to get the energy to keep up that production? What's happening is there's a new thought evolving which goes like this, "Maybe I'm going to make more money producing 17,000 or 18,000 pounds of milk because I'm going to have a lot less problems with these cows. The bottom line means there's going to be more money in my pocket" — net income vs. gross income.

Most conventional farmers have their mind set on a herd of 25,000-pound-producing cows, and then making a lot of money selling the calves to other dairy operations. The problem is that there is no calf crop because they have to replace all their own cows too quickly and there are no heifers left to sell. On paper it works out fine. If they could maintain the herd health and longevity, they would sell a lot of 25,000-pound offsprings for a fair amount of money, but they are using them for replacement instead. So how are you coming out ahead? You're not. It's costing you a lot of money to make that milk from feed alone, and it's costing more in veterinarian bills to make that milk. If you don't have replacements, there goes your profit. It's typical in high producing herds to lose almost half of the herd. The real reason is because they are feeding the herd very high protein (alfalfa at 22 percent), which is dangerous no matter how you raise it. There is a certain amount of that NPN (non-protein nitrogen) molecule that is urea. If you're feeding a cow beyond 17-18 percent protein, particularly if it's conventionally raised, you're feeding urea to cows, according to Dr. McCullough who wrote for *Hoard's Dairyman*. And you can't feed urea to dairy cows and keep them around; it's impossible.

If you were going to sprout the seeds you would want to get the temperature up to a minimum of 62 F, preferably 70 F(+). For soaking, all they do is have the genetic disposition to absorb the water and they are going to swell; that's all you're trying to do is get the water in there. Once they swell, the enzyme inhibitors get moved out. You would not grind the oats before soaking. Remember horses eat whole oats and so can cows. They digest better because they don't have any enzyme inhibitors in them anymore. The enzyme inhibitors do end up in the rinse water, but it's not going to hurt anything because it breaks down. The inhibitors are harmful when they are in the gut, not in the environment. This will work for any animal, and it works well for people too.

Humans don't get very many enzymes, but dogs and cats (pet animals) get even less than humans because, by law, it's required that all pet food be pasteurized. Dogs and cats in the wild eat raw flesh when they make a kill. They eat the viscera (stomach, liver and intestines) first. The meat (muscle) they save for a couple of days later because tryptsin (meat tenderizer) is in the flesh and it breaks down the protein so you can digest it. Butchers can hang hams in the shed because the enzymes kill the bacteria. In fact, enzymes can be used to kill intestinal flu because bacteria and viruses are protein. If you eat protease on an empty stomach you can kill the source of your digestive infection. That's also how you get rid of sports injuries. In fact, I've been asked to make a formula for horses that get banged up in races — what we are using is some protease. This gets into the system and goes to the site of the injury. After an injury you usually get a buildup of scar tissue and protease digests it so that the blood and the lymph can enter and heal the injury. Protease is important for arthritis, stomach infections, or for any kind of injury.

Dyna-min is not an enzyme, but it helps the body synthesize enzymes. The more an animal uses its own digestive enzymes, the less it has to use metabolic enzymes. Think of the body as a reservoir with a finite amount of water in it. It has two spigots — one is the digestive spigot and the other is a metabolic spigot. If you take too much out of the digestive spigot (the rationing man says you're only allowed so much), you won't have as much left for other purposes. You need to have a metabolic enzyme reserve in order to have enzymes for metabolic processes to run your body, to have your organs, lungs and heart operate because everything works on enzymes. So if you forced the body to produce more and more of its enzyme capability for digesting food, it's going to have less to keep you healthy as you get older. This is why you should eat raw foods or foods that are high in enzymes or take enzyme supplements. This is one of the reasons why probiotic factors work well in animals.

Some of these microorganisms, like Lactobacillus, *Aspergilli oryzae,* and the fungal organisms are in the commercially prepared enzymes, particularly amylase which digests starch. Processors grow them and put them through a process where they in effect force them to secrete amylase or protease, which they extract. A lot of organisms in the soil are amylase and protease producers; they have to be, because the soil needs the same digestion that the animal's gut does.

Another reason why Dyna-min works is that it is a highly paramagnetic substance. Again the same California research by the people who analyzed the fecal samples, analyzed the material for parasite egg counts. After 90 days the egg count went down to almost zero on the three types of worms (a tapeworm and two roundworms). Dyna-min is highly paramagnetic, worms are diamagnetic, and they just don't get along together. When you create a paramagnetic environment in the gut, worms just don't survive. One thing that Rudolf Steiner spoke about is when you have certain levels of paramagnetic nutrients in your body, you become a receptor for cosmic energies.

In comparison, Carey Reams talked about the nutrients traveling through the air. I think Steiner maybe elucidated this a little more clearly. He said that from the cosmos (other planets and stars) come wave lengths that correspond to each element on the earth, and if that element is in your body at a healthy enough level, then that wavelength of energy is attracted to that plant leaf or animal body. So, in effect, by nourishing your body with enough of these micronutrients in this form you are able to pull in more cosmic energy, therefore needing less physical energy to sustain life. Dr. Reams stated that if you can get the 20 percent of nutrients from the soil for your plants, the other 80 percent is free and comes from the air. When you take in minerals you can tell that you don't need as much food because you're actually drawing in more of these frequencies.

Generally recommended for cows is putting about 40-50 pounds of Dyna-min in a ton of feed, as well as free-choicing it. Every herd is different. Some herds are so unhealthy and there are so many things wrong with the animals that 40-50 pounds of

Dyna-min is not enough; we can't get enough into the animals. Because cows are so undernourished and so full of junk, they are going to need a much higher level of the product in order to get the problem addressed from the start. In those cases you have cattle eating over a pound a day per animal.

What does that amount to? Normally I start with 40 to 50 pounds and free-choice it. Then watch the cows and let the animals decide how much they want. For a swine herd I cut that in half to about 25 pounds because they get free-choice feed anyway. We go to 40 to 50 pounds for dairy because half of their ration is silage and hay, and we can free-choice it and tell if there is something not right in the ration — either in the quality we're feeding or with something we have eliminated which they need. If you want to control parasites you have to feed at a high enough level. It isn't a parasiticide for animals burdened with worms.

There is a fellow in New York who with his two sons runs a pork and beef operation. They butcher their own all-natural meat animals which they sell in the metropolitan area. People as far away as Texas buy their natural meat. Because he grows an all-natural animal he doesn't use wormers, so he uses Dyna-min. He uses the same method for his swine with about 50 pounds per ton and said that prior to using it he was constantly fighting liver flukes, lung worms, and stomach worms. Once he got to 40 to 50 pounds/ton even the liver fluke and the lung worm disappeared. Obviously Dyna-min isn't going up into the lungs and getting the worms out. It's creating such a paramagnetic envelope around the animals — inside and out — that it's inhospitable for these worms, and they just won't live there.

Mange and lice indicate nutrient deficiencies. When an animal has mange and lice it's because there is a deficiency somewhere in the animals system. Skin diseases are internal diseases with unhealthy tissue outside. What that means is there's poor absorption on the inside or there's either blood congestion or liver congestion. When you have mange and lice often it has to do with the pH of the animal. If your animals experience scours, look for mange or lice afterwards and sometimes you'll see a correlation because of loss of mineral absorption.

Iron is very important for vitamin A absorption. Iron is very important for liver function. Iron and vitamin C are synergistic being absorbed into the blood stream. Iron and vitamin A are both synergistic in the liver. You can't really segregate any of these nutrients because you need everything, but you can't use mega-doses of everything. It is not practical and it's probably not even safe. What you have to look at are some of the key items that reflect the symptom. Then you build a circle around that one or two key items. This in turn usually connects you to some of the interrelated organs or conditions that are affecting the symptoms that you see. Things like mange and lice are usually indicative of either blood impurities, liver malfunctions or non-functions, a vitamin or mineral deficiency because of deficiencies in the diet, or because of malabsorption in the gut and then we start talking enzymes again.

My guess is that in most hog operations the enzyme system, as it was meant to be, is not there. Hogs, as you know, are not raised under the most ideal conditions because of management necessities. When you look at the oxygen and sunlight factors, you realize there is a problem. What is a hog good at doing? Rooting! That's what it's made for and it can't do that on concrete. That's why if you put Dyna-min in front of hogs they go nuts for it; they need it. Hogs are designed to be "plows." They ingest a lot of topsoil in the process of rooting for grubs, earthworms, rodents, mast and roots. Topsoil is a very rich source of soil-based probiotics that produce enzymes for proper digestion and naturally occurring antibiotics that prevent diseases. Hogs on concrete are susceptible to infections and so are force-fed synthetic antibiotics.

I once read a magazine article about a farmer in Nebraska who basically is hog ranching. He leaves the hogs outdoors, corrals them in, and moves them around. Basically he is following a kind of rotational grazing system with the hogs. He lets that area rest for a while and eventually it settles down, cleanses itself, and becomes regrown sod. These animals are living in a native, indigenous environment and they are healthy. He has solved a lot of health problems by using this rotational grazing system because they get the 20 percent oxygen from fresh air that they need and the sunlight which gives them much more usable vitamin D than just having to feed it to them.

One more comment on free-choicing. If you see your animals eating Dyna-min (or other free choice supplements) like they're inhaling it, what does that indicate to you? Animals attempt to eat what they need. Animals will eat what they need. They won't eat something if they don't need it. If they are overeating it then that is an indication to you that there is something wrong. They must desperately need the stuff. A lot of farmers say, "I'm not putting that stuff (Dyna-min, kelp, etc.) out there because they're eating it like slop "chop" in just a few days and they will eat it all up." Well, they must need it then. Stress will kill a lot of the microflora in the gut and the gut will not operate efficiently. If you go into some dairy operations you'll see where the animals walk around a lot, particularly around fences. These are areas where they've eaten the clay. That's either a sign of acidosis, runaway ammonia, toxic feed, toxic water, or something else — mineral deficiency, etc., Dyna-min and certain clays are rich in silica, an important mineral in building bones, teeth, hoof, hair coat, cartilage, etc. and is found in certain grasses and herbs. Together with a calcium-rich plant such as alfalfa or comfrey, you have the raw materials for skeletal formation. Comfrey has a 2:1 calcium-to-phosphorous ratio. That's why comfrey has a nickname of "knit-bone." Some of these clays, like Dyna-min, are a reservoir of bone building elements, namely calcium, magnesium, silica, boron, zinc, strontium and manganese. They are usually short on phosphorous, so supplying phosphorous as fodder and/or mineral supplement helps complete the mineral requirements.

Some of the original plant materials that Dyna-min is high in are horsetail, shade grass and bottle brush. Bottle brush is an herb that the Indians used to use as a urinary antiseptic, but now it's used in many of herbal formulas for building bone and teeth and rebuilding cartilage damage. The reason why it's so effective is they usually use it with something like comfrey. It's a great plant to grow, except that it's a high-moisture plant. Because it's a wide leaf, you can't dry it down as easily as alfalfa. It has a 2:1 calcium ratio instead of 4:1 like alfalfa. It's also one of the few land plants that has vitamin B-12 in it. If you want good calcific absorption, use horsetail, silica, and other things including boron and manganese. If you look at an analysis of Dyna-min, you'll see that it has 12 percent silica. Silica could be window glass or beach sand or it could be the kind of silica you find in Dyna-min or horsetail — a silica which is biologically available to the bloodstream. When it gets in the bloodstream, it's synergistic with calcium so you can build teeth and bones and you can provide all the functions of calcium.

I think you should force feed about 4 ounces of clay a day. I think that's a good level because it's not excessive and you're not pushing the animals at it. If you free-choice it and they take more than that, then watch the clock and see why they are eating so much more. Sometimes herds can be a little finicky about Dyna-min additives such as clay or humates; so if they won't eat the grain after you've put in these supplements, try smaller amounts until they get used to it. When animals are free-choicing you should be looking for toxicity, low calcium, poor calcium absorption, mineral imbalance, low food enzyme levels, nitrogen or ammonia excess, and acidosis.

Ammonia is there because of the way we grow foods and what we feed to the dairy cattle. High-protein alfalfa or excessive soluble protein diets can create ammonia in the bloodstream which is acknowledged as a poison by the animal. It goes directly to the liver, converts to urea, and is sent back out to the bloodstream. The kidney can pick up the urea and get rid of it as uric acid. That's a lot of energy that could have been used to make milk or grow an animal. It could also cause an abortion. Urea in the bloodstream feeds bacteria that you don't want, it knocks the dickens out of the white cells and it suppresses the immune system while it feeds the bugs you're trying to get rid of.

Ammonia absorption is one of the reasons why Dyna-min and certain clays work so well.

The California Department of Agriculture acknowledges that Dyna-min has a matrix of micronutrients and that this product is a nutritive, natural occurring mineral. It is classified by other states as montmorillonite clay.

The above discussion amplifies the serious problems in livestock that result from our farming methods, the quality of our feedstuffs and our attempts to compensate for the situation.

Chapter 20

Tillage

Ideally, soil would have 5 percent or more organic matter, 25 percent air space, 25 percent water and 45 percent mineral content. The ideal soil would have a crumbly structure, even to a depth of 10 to 12 inches with an aroma resembling a fresh virgin forest. Soil particles would be flocculated such that air and moisture flowed easily. Having these characteristics, the soil would be identified as having ideal tilth.

Soil Tilth

If the soil tilth is satisfactory, minimal tillage will be needed. Proper tilth means the soil is loose, crumbly to the touch, drains well, decomposes well, has plenty of air space, and is soft to the touch.

If the tilth is good, plantings may be done with minimal fitting of the soil. Even in this day of chemical control, many farmers find they have better weed control with cultivation. The benefit of air introduced into the soil is often an unexpected plus. The air assists the development of root mass and supplies microbial life with needed oxygen.

Although virgin or untouched soils will differ from this description, the farmer has certainly played his role in reversing or altering this ideal characteristic. Working the soil under wet conditions is one cause for soil compaction. Another source is operating excessively heavy equipment. Additional problems arise when you continually plow or till at the same depth, fail to manage residue, kill off or otherwise interfere with biological decay (which decreases organic and humus levels), use excessive tillage passes, and over-apply potash and magnesium at the expense of calcium.

It has been generally thought that larger implements could be used safely as long as large ties and/or dual ties were used. However, even this concept may be in error. The March, 1985 issue of *The New Farm* contained an article describing the effect heavy equipment has on soils of varying moisture levels. Quoting information from the Ohio Cooperative Extension Service, it concludes, "That it is total weight, not pressure per square inch, that causes the really troublesome kind of compaction." A 13- by 30-inch tire carrying 2,200 pounds compacts wet soil almost twice the depth that a 7- by 24-inch tire carrying 660 pounds does. "With big, modern machinery, we're getting soil compaction as much as 27 inches deep when farmers work ground wet," says Dave Kopcak, district conservationist in Wyandot County, Ohio. "That's not only too deep for plows or frost to break up, but questionable for even the biggest subsoilers."

Besides preparing a seed bed, tillage can aerate, break soil crust, and control weed growth. In addition, it can improve the soil's magnetic flow and release energy. Excess tillage can destroy humus and soil structure by adding excessive air to the soil. The additional air (oxygen) speeds decay of the organic matter, especially in southern and southwestern soils. Tillage requirements relate to several factors including lay of the land, water table, soil type, and fertility practices. Soil types can be obtained from your local ASCS office. Improper fertility factors often can be assessed through power requirements needed for tillage (equipment pulls hard) and from weed patterns.

Fertility relates to compaction as discussed in Chapter 11 under potassium. Excessive emphasis on potassium applications at the expense of calcium (lime) applications have damaged the soil structure. As the potash usage has increased, the potassium ions have replaced the calcium ions on the clay colloid. Magnesium ions will do the same. As calcium is increasingly replaced or driven off the clay, the soil structure begins to change. The clay soil structure collapses becoming more sticky in the spring and hard and compacted in the summer. High-calcium lime and soil conditioners can reverse this process by flocculating the soil.

A flocculated soil means the soil is loosened up and has more tilth. Such a field will plow and otherwise work better than soil lacking flocculation. Failure to add calcium means the calcium ion is not available to perform the flocculation function required in soils characterized by soil tilth. Soil compaction is related to other variables besides fertility. Working the soil under too wet conditions is a significant cause for compaction. It also contributes to soil clodding. This practice compacts soil, pressing out the air spaces; once the soil dries out, it retains its new shape. There is one rule that is sometimes ignored to the detriment of the farmer and his soil: *Don't work the soil too wet.* It is better to plant two weeks late than to "mud it in" on time and lose several weeks because of poor growth in a bad soil. This applies to any field operation.

Compaction can also be related to misuse or overuse of tillage implements. Improper plow point or disk adjustment can cause a plowpan to develop. This, as all forms of compaction, is damaging because it prevents moisture from moving up and down in the soil through capillary action. Many farmers find a lake has developed in a low-lying area of a field. This usually occurs after several years and can be directly related to compaction. In Chapter 22, we will discuss how to eliminate this ponding effect using non-tillage means.

Tillage is only one way to work with this problem condition. Residue management is increasingly becoming a tillage priority. It is now apparent that crop residue is a vital key to soil tilth. Incorporating residue provides soil bacteria vital nutrients otherwise lost to the sun and elements. Tillage should not be a perpetual patch job. There must be a system of tillage and a system of fertility which complement each other. Any change in tillage systems which ignores soil fertility will probably never achieve the optimum of crop yield, quality, or profit desired. It may be necessary to loosen tight soils or break hardpans. If the soil magnesium level is too high relative to the calcium level, the desired improvement in soil structure and aeration will probably not be permanent. Deep tillage may need to be repeated because the cause of compaction has not been eliminated and the mechanical opening will revert to a compacted state. Proper calcium, and magnesium levels will do much to reduce horsepower and fuel requirements in tillage operations. This will help in each tillage operation and may reduce the total number of such operations required.

The Chisel Plow

Does subsoiling pay? Compacted soil quickly closes back together, and hardpans seal the soil, interrupted only where the subsoiler has passed. Examining the soil reveals the top 10 to 12 tillable inches often has one or two tillage layers. This is caused by a variety of soil particle sizes tightly packed together and having a higher moisture level. This soil condition requires the farmer to buy bigger tractors to usually pull smaller plows. The plowpan may be 2 to 5 inches thick. Roots often will not penetrate this layer and moisture will not pass through it. The subsoil below this hardpan is usually mellow, crumbles easily and the soil particles are coarse and usually uniform in size. Chisel plows are commonly utilized in these situations.

It's important to evaluate the results obtained from breaking hardpans. Obviously, the point needs to be lower than the plowpan itself. Yet, other variables, such as point spacings, are also important. Is the soil simply being torn through with only a slight fracturing between points? Is the deeper depth of 16 inches merely using up fuel and traction? Is the entire section being lifted? A central Illinois implement manufacturer markets a chisel plow which lifts and fractures the soil uniformly between points. Each manufacturer has his own approach to breaking the hard pan. Review the results of their work before purchasing.

When thinking about chisel plows, choose the correct shovels for the job. Most machines will accommodate the 2-inch, 3-inch or 4-inch twist shovels. Usually, the deeper the tillage, the more narrow the shovel. In late spring, the narrower shovel will expose less moisture to be lost to the atmosphere. The wider shovels require more power, especially at greater depths. Seven three-inch twist shovels pull harder than nine two-inch spikes in the same field. It is important to know what you wish to accomplish in a given situation.

Residue incorporation is also a function of chisel tillage. In some situations, the chisel plow and coulter chisel do well because they mix the residue near the surface of the soil. This may be especially helpful if a soil has only a 3 or 4 inch aerobic zone. To moldboard plow residue 8 to 10 inches deep in this soil condition is to almost guarantee that there will be little decay system and no new humus formed. The aerobic bacteria will be buried below the oxygen level while the anaerobic bacteria will be left on top exposed to the air. The residue will ferment, producing an alcohol or aldehyde. These substances kill off the aerobic bacteria and preserve the trash.

The Moldboard Plow

In spite of the negative aspects involving erosion and power requirements, the moldboard plow can be beneficial. In the soil, some nutrients tend to rise while calcium and others tend to move downward. A soil left undisturbed will stabilize from the top down in the following layers: carbon, magnesium, phosphate, potash, sulfur, aluminum, manganese and calcium.

At times, it is very beneficial to invert this system with the moldboard plow. This inversion will bring the heavy nutrients to the top and move the lighter nutrients down into the soil. It may be best to remove the trash or cover boards from the plow. This will leave some residue exposed to the soil surface for snow and moisture retention. It will also help wick air movement into the lower part of the furrow. When using primary tillage implements, such as moldboard or chisel plows, it might be helpful to open the soil surface first with a disk or field cultivator. This is particularly true when working a sod field. If the soil surface is broken first, the soil tends to flow better with the plow, leaving a smoother surface.

Farmers have talked about plowing weed seeds down to avoid weed problems. Plowing will help if the soil environment is suitable for the seed to decay. It probably doesn't work too well in some fields because some of the more harmful fertilizers tend to preserve the seed for growth at a later time. Weed seeds which are plowed down will be plowed up next year to cause problems. Is the soil managed to decay or preserve the weed seeds?

Hardpan

Most tillage approaches can produce a plowpan or hardpan. The moldboard plow carries much weight on a very narrow edge of the plow share. In wet conditions, the soil below the plow share will smear. As it dries, it will seal stopping water and air movement. Disks, chisel plows, field cultivators, and subsoilers can all contribute to hardpans even in sandy soils. As the soil is tilled, the small particles settle. When tillage is continued at the same depth, the particles settle just below the tilled level. These small particles keep filling the pore spaces until a hardpan is formed. This can be just as bad a hard pan as that caused by plowing when the soil is too wet.

This hardpan stops water movement up and down. Plant roots can't penetrate and water stress is the result. It has been estimated that if a non-compacted soil will yield 185 bushels/acre, it will drop to 140 bushels/acre on moderately-compacted soil and down to 70 bushels/acre on compacted soil.

Donald Schriefer of DeMotte, Indiana, is known for his consulting work pertaining to eliminating the hazards of the hardpan. In his book, *From the Soil Up*, Schriefer showed in both pictures and diagrams how corn height and yield is stunted from hardpans. His solution is to break this hardpan with an anhydrous knife at cultivation. He has progressed in his thinking from placing the knife behind the tires to placing it next to the row. Once opened, the roots are then able to follow the opening deep into the subsoil to gather moisture and minerals.

Ridge Till

Ridge till is a system supposedly well suited to growing continuous row crops. As the crop is cultivated the last time, soil is thrown or ridged around the growing plants. Next spring, at planting, the top of the ridge is scraped off and the seed is planted on a leveled area on the ridge. The ridges are rebuilt at the second cultivation. When properly managed, wheel travel is limited to only certain rows. This allows the soil to remain loose between the rows encouraging better root growth. The residue reduces erosion between the ridges. Cultivation mixes this residue into the soil. The last cultivation spreads the remaining residue around the base of the plants. Some farmers build the ridges every spring and break them down every fall, permitting better mixing of the organic matter. The soil on top of the ridges warms earlier in the spring plus remains elevated above the field height. This allows for better drainage and permits earlier planting and crop growth.

Ridge till also has problems. A major problem pertains to residue management. Trash is often left lying on the soil surface with little effort given to incorporation into the soil. Residue left in this manner will actually rust, similarly to rusting equipment with the beneficial carbon being lost to the atmosphere. Residue must be incorporated so that beneficial soil microorganisms can reduce it to humus.

Secondly, although ridge till was envisioned as a system which eliminates soil compaction, a walk in the field often reveals a different story. One of the authors walked several ridge-tilled fields in the summer of 1989 using a penetrometer. This instrument measures the pressure needed to push a 1/4-inch rod into the soil. Those checks taken in the wheel wells were often easier to push than in the ridge itself.

No Till

A few years ago, the front cover of the *Michigan Farmer* showed a picture of erosion in a no-till wheat field. The question was asked, "Why?" The following paragraphs will attempt to suggest the answer to that "Why?" For this discussion, no till is limited to the practice of weed control by chemicals with the crop planted in soil which is not tilled. Weeds are controlled strictly through chemical means.

In no till, the planter is equipped with some type of disk or opener to loosen the soil in the row where the seed and fertilizer are applied. No other tillage is used. At least one of the herbicides used is a burn down material. It is intended to kill everything green. It is usually applied before or just after planting, but before the crop emerges from the soil. The weeds die and begin to decay. Looking back at the residue paragraphs in this chapter, remember that carbon and nitrogen tend to rise in the soil. Looking at the periodic chart of elements in any chemistry book, one finds that carbon and nitrogen are lighter in atomic weight than oxygen. As the plant decays, a portion of these lighter elements are carried into the atmosphere.

There is no means for mixing the crop residue into the soil for humus formation with no till. If the residue were cut and laid on the soil surface, the earthworms could carry some organic matter and minerals down into the soil. Herbicides on the crop residues, however, may disperse the earthworms. The soil in a no-tilled field is usually much harder than in a good, biologically farmed soil. In analyzing long-term no-till fields, the humus and soil porosity properties often are lacking. A farmer in Illinois has observed the tillage changes of the past few decades and summed them in the following phrase: "Tillage has gone from the moldboard plow to the chisel plow to no till to no farm."

Secondary Tillage

Some of the new field finishing implements are intended to do all secondary tillage in one pass. This is a good idea in good soil conditions. If the soil moisture content is correct and the soil flows over the shovels and blades, these implements are helpful. As the soil moisture content increases, they appear to do a less satisfactory job. Anytime it is necessary to *beat* the soil into condition, something is wrong. Sometimes two passes with a more simple implement such as a disk or field cultivator will do a better job. It is best if a few hours of drying time occurs between passes. A larger field cultivator — relative to available horsepower — run at a more shallow depth may allow two passes in about the same time as a small cultivator run too deeply.

The benefit of not abusing the soil will offset the extra fuel cost. At least one secondary tillage manufacturer claims a single pass produced identical weed kill compared to the more conventional method of two chemical incorporation passes using a disc harrow and then a field cultivator.

Other than a little starter fertilizer, preplant fertilizers should be mixed in the top layers of the soil. The idea of plowing down fertilizers (including lime) has one major disadvantage. When plowed ten inches deep, the minerals may leach another four inches making it nearly impossible to reclaim them by tillage. The only hope is for deep-rooted crops to try to salvage wasted fertilizer dollars. Those minerals leach fast enough without our help. Some crops, like carrots, may be able to make better use of deep nutrients. Mixing the fertilizer in the top 2 or 3 inches creates an ideal location for microbial decomposition. As the aerobic zone increases, the depth can increase.

The Disk

Robert White, formerly with Michigan State University Agricultural Engineering Department, says the disk will destroy soil structure more readily than any other tillage implement except possibly the rototiller. The disk blades cut the small fibers and fine root hairs which hold soil particles apart. The curve of the blades and the angle of the gangs tends to pack the soil. Especially when the soil is wet, a disk is a poor choice of tillage implements unless operated very shallow to break the crust and allow the soil to dry. The disk, either offset or tandem, should be used only when necessary to cut residue or weed growth.

A personal experience from Jay McCaman: He was preparing a field for fall grain using a chisel plow and a field cultivator. One large area of sod was not breaking down as desired. "I used a disk to work this area and in the process made two passes across the field with the disk. It rained before I worked the field again. On the next tillage operation, I went across the two disk passes. The remainder of the field was dry enough to work, but the tractor wheels spun nearly every time I crossed those two disk passes."

Cultivation

For row crop cultivation, stagger the shovel placement next to the row. The soil from the lead shovel will tend to tip the weed over while the second shovel covers it. When the shovels are directly opposite each other, there is a tendency to prop up the weed as well as the crop. This principle applies to disk hillers also.

Deep cultivation too near the row can prune crop roots. This is more damaging on larger plants with large root systems. This pruning may result in loss of moisture and nutrient content by the plant.

A California manufacturer builds a combination torsion weeder and spider to remove weeds from the row. These units can be mounted on regular cultivators. The spiders work close to the row, throwing soil away from the row and covering weeds. The fingers on the spiders are staggered so that the edge of soil near the row is not smooth. The torsion springs are set to rub against this uneven edge which makes the springs vibrate. This vibrating action causes the soil in the row to crumble, upsetting small weeds. Again, early cultivation is crucial.

Weed Control

Tillage, including row crop cultivation, can control weeds and help turn them into humus. Quackgrass and other rhizome-type grasses can be killed when the roots are exposed to radiation from the sun. A straight shovel from a field cultivator will shake out many roots for a better kill. If the root pieces are allowed to start growing, the problem could become worse. It is possible to kill quackgrass with a disk, but the soil structure will probably suffer from the many passes required. A moldboard plow may bury the roots deep, leaving a deeper root system to kill by other methods. As weeds become larger, a cultivator with a straight shovel may jump around a well-rooted plant. In such situations, a good sweep or disk can be used to cut the weed below ground level.

A principle to remember: The smaller the weed, the more easily it is killed by tillage. Most weeds germinate within the top 1/2 inch of soil. The best time to kill weeds is when they are "in the white" which means before leaves turn green. The energy reserves of roots, seeds or rhizomes are low just before the leaves turn green. The green in the leaf is chlorophyll at work producing energy for the roots and other plant parts. Preventing chlorophyll formation is a big step in weed control. This helps explain why multiple tillage passes spaced a few days apart can control quackgrass and other weeds. A rotary hoe is good for both removing young weeds as well as for loosening up the soil. Deep tillage can bring more weed seeds to the surface to germinate. Many times it is only necessary to disturb the soil surface in order to control small weeds. Various harrows and drags do this well. Night tillage will obviously eliminate the sunlight needed for seed germination.

Although most weed seeds germinate in the top 1/2 inch less of soil, one exception to this is velvetleaf which can germinate at least as deep as 3 inches. Velvetleaf likes an anaerobic environment. If the crop is planted deep enough, it is possible to work the surface of the soil after planting and eliminate many weeds after they germinate, but before the crop emerges. The crop will emerge a few more days ahead of the next weeds to germinate. An Australian manufacturer has developed a spring-tooth chisel plow which, it claims, eliminates subsequent weed growth. Dr. Philip Callahan discusses this invention and provides an explanation as to why it works in his book, *Exploring the Spectrum*.

One farmer used the Feurstine harrow very successfully. He went over his field five days after planting at a 30-degree angle to the row. By working on the angle he eliminated the small ridges which the planter left along side of the rows. This left a level surface for the cultivator. There were no large clumps for the cultivator to push over onto the young plants.

Thanks to Jay McCaman and Thomas Besecker for contributing to this chapter.

Chapter 21
Field Evaluation

One of the best ways to confirm the ideas and concepts in this handbook is to go out in your fields and see for yourself what is really going on. Driving by your corn field in a pickup truck may make you believe you have a great crop. If you are selling corn to the market and you actually end up harvesting "X" number of bushels which gives you what you consider a good profit, then maybe you're right in saying it was a great crop. But that's looking from a limited viewpoint.

You can also look at genetic potential and see how close your efforts came to that. You can look at soil or water damage you might have caused by nitrates, toxic pesticides, and erosion. You might examine feed value and determine if your corn really is something you can be proud of in terms of what it will provide for livestock or people. Looking at these broader concerns just might make you realize that the great corn crop really could stand some improvement in ways that aren't as obvious as just going over the scale or looking out the pickup window.

Observational Skills

Walking into a field and thinking about the atmosphere you experience instead of just looking at corn stalks may be something you have never tried. Get a feel for the nature of things. Are there birds flying around? Are there insects — good or bad — in the field? Do you see any animal tracks or droppings? Do you smell good earth? Does the soil look hard, dead, sterile, and free of any other growth except corn? Can you see cracks and erosion? There is an obvious difference, if you let yourself sense it, between an anhydrous, herbicide, fungicide, and pesticide treated field and a living field which has little or none of these products used on it.

What about weed patterns? Are there lambsquarter and redroot pigweed indicating improper phosphorus to potassium ratios? Are there nightshade and velvetleaf indicating soil degradation? Are there sour grasses, e.g., foxtail, and Johnson grass indicating available calcium shortages or calcium:magnesium imbalances? Now, let's look at the corn stalks. (The observation and principles suggested in our example of evaluating a corn field will apply to all crops. Significant variations for other crops will be noted.) Do you see purple discoloration on the corn stalks indicating a phosphorus deficiency? Do you see firing or veinal discolorations on the leaves indicating a variety of excesses or deficiencies? Is the corn a good vegetative green or a deep, dark green indicating a nitrogen excess? Are there several layers of brace roots indicating plugged nodes which are preventing adequate uptake and transport of nutrients?

Plant Evaluation

Now let's examine the corn plant. You can start with a brix reading, using a refractometer. Choose a particular leaf position if you are going to check several stalks in the same field or make comparisons with other fields. Even the parts of the leaf may have different readings, so you may want to establish a spot such as the midsection of the leaf opposite the main ear (the leaf one node above the node where the ear originates). The range for corn stalks can be from 4 to 20 brix, but exceptions in either direction can occur. Taking into account all the technology associated with brix readings and their variations, try to see what the reading is telling you. If the reading is low, below 8 to 10, and the corn looks very healthy with little or no insect or disease damage to ear or stalk, you may be looking at a temporary low brix condition. Check again in a few days.

Raising the brix later in the season is too late to make up for the previous low brix condition which may have contributed to failure to establish a large number of double rows of kernels, fill out the ends of the ears, or to prevent possible ear or stalk insect or disease damage at an earlier stage. If the corn shows obvious integrity and the reading is reasonably high, 10 to 14, you may be moving in the right direction. Cut the stalk off about 18 inches above the ground and dig out the main root mass (save both pieces). Examine the root mass for general configuration, fine root hair development, and directions of travel in the soil. Distorted root configuration may lead to discovery of toxic spots in the soil. Sometimes these can be identified as in improperly decayed vegetation that the roots are avoiding.

Next, try gently pulling on a medium-size root to see if the root bark will separate and slip off easily like a stocking. This would indicate weakness caused by excessive salts in relation to carbohydrates and humus and could provide a situation where nematodes could easily penetrate.

Observe the cut off end of the stalk for the closeness of its shape to a circle. The more elliptical or oval it is, the less calcium was or is available for your corn crop. Using a sharp knife, start at the top of the cut off stalk and split it all the way down through the root tip and root mass, exposing the inner surface of the stalk and root. Now you can see the real health and integrity of your corn. The root tip should be white and structurally intact. In most cases, it is discolored and even rotted. Notice the nodes where the brace roots come out. They will appear discolored, darkest at the lower end, and lightest above the last set and above.

Never taste a discolored node as you may be ingesting a toxic dose of whatever your plant is trying to filter out of the plant juices. Tasting the clean, clear white pith from higher up can give you a relative palatability check for your livestock. The sweet taste in a healthy corn plant is comparable to chewing gum and the sweetness can be measured by a refractometer.

When you have finished looking at the stalk integrity, you are ready for the upper part of the corn stalk that you cut off previously. You can check the genetic potential of the corn to produce ears in comparison to what you currently have showing on the stalk. Your current accomplishment may be one good ear and possibly a second in various stages of development.

Start at any node. Carefully peel the leaf back to expose the indented side of the stalk. As you work down the indentation, you will discover an ear of corn in some stage of development. Usually you will find what appears to be corn husks waiting for a corn ear. Working right down to the point of connection, you should find a tiny ear, one or two inches long with clearly identifiable number of rows and kernels. Do this at each node, and you will usually find five to seven ears that you are not going to harvest. Obviously, you haven't provided the proper soil structure, soil fertility, and moisture to allow the corn to develop those ears. This observation may be ego deflating, but use it as an educational and motivational experience instead: *What can I do to tap the genetics on my farm?*

The above procedures would apply to all crops except trees or bushes which would not be economical to sacrifice for a field demonstration. If the plant is a legume, such as alfalfa, clover, soybeans, peas, or dry beans, root examination should include nodule observation. If you find any of the pinhead or larger size round nodules, check the inside for pinkness which would show proper nitrogen fixation. Frequently, you'll find few, if any, nodules on chemically farmed legumes. Rather than nodulation nitrogen, much of the so-called crop response of a crop following plowed down legumes will come from the nitrogen content of the plant itself as well as the plant growth hormone effect, particularly from alfalfa.

In perennials such as strawberries, sacrificing a plant or two is no problem. Look for new roots coming off the new crown in relation to degradation of the old crown. Sometimes the latter precedes the former, seriously weakening the ability of the top to get nutrition.

In alfalfa, grasses and grains, check the hollowness of the stem cross-section. The optimum state is total filling-in of the stem with tissue would be indicative of calcium and boron availability. Hollow or less than full stems mean that you've got some soil work to do.

Cotton irregularities are particularly observable at maturity. A large number of green bolls present, along with the mature cotton, indicates a failure to move into the dormancy stage. Excessive nitrate nitrogen could be a factor as well as a variety of mineral shortages or imbalances.

Tree Evaluation

Fruit and nut tree observations get a little more complicated, but the principles remain. Leaf color, insect damage, and physiological disorders are all easily observed. The length of new growth observed in late season is relative to the type and breed of fruit or nut. The same relativity applies to the ratio of fruit buds or vegetative buds.

One obvious example of the tree being out of synch is when spring checks show freeze-killed or weakened vegetative shoots that grew the previous fall. Sometimes a few fruit buds develop in the fall and may actually have a blossom present. The normal tree process should be smooth and consistent throughout the season and natural dormancy should occur in the fall. This means that the tree should have a normal June drop to adapt to the nutrition it senses as being available. It should not proceed to have cracking or splitting of fruit or dropping prior to harvest.

The necessity of stickers, shapers, and color enhancers are indications of nutritional (energy) deficiencies or imbalances. Bark and limb integrity should be observed. Split bark, broken limbs, and brittle wood may be copper deficiencies. Poor pollen production and unattractiveness to bees is certainly a growing problem and is directly related to brix content.

Getting good production only every other year may just be an inadequate supply of energy due to not having the right fertility available to produce the needed energy. Energy can't be stored in the roots if it's used up faster than it is replaced or not produced in the first place. The list of fruits, berries, vegetables, nuts, legumes and field crops produced is almost endless.

Therefore, in summarizing, you need to look for two basic problems besides the obvious basic diseases, insects or conditions. The first area is a failure to follow physiological steps of growing, fruiting, and maturing. Alfalfa, lettuce or spinach that goes to blossom or bolts early indicates a fertility imbalance situation that may be worsened by weather extremes.

The second concept is that of physiological disorders (conditions for which no diseases or insect vector can be found). Any time you see cellular structural degradation or distortion of a plant, the plant is not operating on full efficiency. Conditions — hollow heart in cauliflower, split pit in peaches, undeveloped nut meats, etc. — are all signs of the nutritionally imperfect situation. Plant tissue does not follow the thermodynamic laws of mechanical devices, but maintains efficiency (no loss of energy) the harder and faster it grows.

Chapter 22

How to Recover a Field

Stop Using Destructive Products

The first step in the recovery of a field or soil is to know what products to stop using. By discontinuing the use of destructive products you will begin to see the soil recover on its own. If you will stop using anhydrous ammonia, triple superphosphate (0-46-0), muriate of potash (0-0-60), and (usually) dolomite lime, you will begin to see your land recover. You will stop destroying bacteria once the chloride washes out, and you will stop tying up calcium and phosphate once the 0-46-0 is discontinued.

Magnesium levels in excess of 15 percent on a CEC soil test can create sticky spring soils and compacted summer soils. Therefore, it is generally best to avoid the use of dolomite lime. Correct magnesium deficiency problems using epsom salts or chelated forms of magnesium either broadcast or foliar applied. When soils are low in magnesium and in low-CEC soils, consider using dolomite in the prill-lime form. This form is more capable of being applied at reduced rates which will allow you to control the magnesium being applied.

Making Changes

Farmers considering changing their fertility approach need basically three things: a model to follow, a consultant to discuss programs and progress with, and a good product base to draw from.

Settling on a model to follow is about as difficult as settling on a vitamin program to follow. Conventional wisdom says we don't need vitamins, and the experts in the vitamin field all disagree with each other. It is just as confusing in the field of alternative agronomy, but don't let that prevent you from starting a program.

There are many ways to accomplish a change. Some may be better than others, yet all may work relatively well. The same is true for programs and consultants. Basic philosophical differences are good, even healthy, because they keep us alert to finding the best approach. The main point is to keep going, continue to look for help until you feel comfortable with your approach and *don't quit!*

It seems risky to venture into something you are unfamiliar with especially if you don't have neighbors or other supporters who are also experimenting and could help you solve problems (challenges) you encounter. But difficult as it may be, the time is quickly coming when many rescue chemicals will be classified as either illegal or highly restricted. It's imperative that farmers make preparations now.

Initially, you want to record information about the field(s) you plan to work. Start on a small scale and work up. Don't bite off more than you can chew. Record the weed patterns and consult ecological agriculture reference books to see what soil conditions the weeds are indicating. Send off soil samples to labs that offer LaMotte and electronic scanner tests, in addition to obtaining a CEC test. Review calcium levels and phosphate/potash ratios and compare these figures with the weed populations observed.

Review past years' fertility programs. When was lime last applied? What kind of lime was used? What type of fertilizer was used? What kinds of chemicals have been used? How is the soil microbial life (do you plow up last year's corn stalks)? Do you have standing water? Do you have areas which won't grow anything?

Fall Programming

Soil Flush

It is best to begin a program in the fall. This timing gives the soil several months to change prior to growing another crop. If toxic chemicals or salt residues are present, it would be wise to consider a soil *flush*. This consists of using a wetting-type product to literally flush out the hydrocarbon residues and salts into soil layers below the root zones. As much as two quarts per acre might be indicated.

Many farmers have been convinced of the value of using wetting agents on an ongoing basis. The authors' experience is that within a few years, especially if larger amounts (more than two quarts per acre) are used, the original problems of compaction will return and nutrient stripping can result in imbalances. The soil flush concept has a specific purpose. Once this purpose is reached, it is no longer appropriate to continue.

Wetting agents and soil conditioners come in several types. Some use linear alkali oxalates. This type works by making water *wetter*. They reduce surface tension, thereby allowing the water to more easily penetrate the soil. Other types contain ammonium laureth sulfate. Dr. Reams used this second product and believed it flocculated the soil by spreading apart soil particles and causing air spaces to open. Both products will usually feed soil microbes. When using either type product, it is best to apply when soil moisture is good or when a rain or irrigation is imminent. These types of products help reduce ponding or poor drainage which may be caused by management practices that produce compact, salty soils.

Evaluate Mineralization Levels

Fall is the best season to consider a lime and phosphate program. High-calcium lime along with colloidal (soft rock) phosphate would be preferred. Gypsum would also be indicated if high pH levels, heavy clay, and/or compaction is present. High-cal lime traditionally would be used at 1 to 2 tons per acre with gypsum used at 500 pounds per acre. Apply colloidal phosphate at 200 to 500 pounds per acre. These amounts apply in the north with increases acceptable as you go south toward longer seasons and warmer temperatures. When applying these materials, apply the colloidal phosphate first and then apply the lime. This combination will create an energy (heat) response which is sufficient to kill surface weed seeds.

If applying in the spring, it is best to wait ten to 14 days before planting as seed germination may be affected. These two products have a tremendous benefit to soil, microbial and plant life. Although this level of lime application may appear on the low side, new evidence suggests that even more appropriate applications might consist in applying as little as 150 to 250 pounds of a granulated high-cal lime more often than once per year. Some farmers have found such applications following each alfalfa cutting have greatly affected crop yield and quality. Smaller amounts may be more palatable to soil microorganisms. The lime will begin to expand the soil by separating the microscopic layers of clay which helps create air spaces for bacterial growth as well as provide calcium energies to the bacteria.

The colloidal phosphate will also assist in soil flocculation, create an excellent base for microbial growth, and the two in combination will establish a strong magnetic (energy) field. If colloidal phosphate is unavailable in the amount you need, or if it is cost prohibitive, consider 10-34-0 or 11-52-0 as interim measures to help stimulate basic soil phosphates.

A farmer in Kansas applied colloidal phosphate. "I can't afford 500 pounds/acre, but I will apply 300 pounds and include my gullies," he said. Four weeks later he called the consultant to report. "Shortly after applying the colloidal phosphate, I planted wheat. The same day my neighbor planted wheat in an adjoining field. Last week we had a terrific wind storm and soil blew all over. After the storm, I noticed that my fields had not blown. In fact, I barely had gullies any more. Yesterday, my neighbor replanted his wheat, but mine does not need replanting." This man has experienced what we are trying to accomplish, namely establish an energy pattern over the surface of the field. This energy can eliminate blowing and help attract minerals to it.

Although colloidal phosphate is important to use for its benefits of colloidal trace minerals and colloidal phosphate, it may not be the most economical source for basic phosphate building in soil. Black hard rock phosphate has a fine, granular texture that is not dusty and contains 30+ percent P_2O_5. It should be considered for soils with very low phosphate levels. You can still include colloidal phosphate later.

Evaluate Sulfur Levels

When sulfur levels are low, consider the use of a good dry or liquid sulfur product such as ammonium sulfate or ammonium thiosulfate. The sulfur will work with the calcium and phosphate in creating soil energies. Sulfur is commonly used to release calcium and make it available to the plant. It is also a requisite for the efficient use of applied nitrogens for making many amino acids and complete proteins.

Biologicals (Soil Microbes)

Soil mineralization is a requisite for good microbial activity — especially a proper balance of calcium, magnesium, phosphate and potash. When calcium levels are too low, most microbial products will not work because microbes need calcium to thrive. Calcium levels must be at least above 60 percent of base saturation to insure microbe survival.

Most chemicals were supposedly built to be decomposed naturally by soil microbes. The problem is that they have created energy patterns in the soil which are not conducive to beneficial microbial life. Most toxic materials are applied according to soil humus levels which are directly related to microbial functioning. The more humus present, the more chemicals you need to use to get by the buffering effect of soils. As the humus levels are depleted, you should be able to use fewer chemicals to get the same effect.

Adequate microbial life is a strong defense against the accumulation of toxins in the soil as the humus produced serves as a buffer against both salts and toxins.

Soil bacteria require air, water, food and shelter, just like any living organism. They usually have plenty of nitrogen for their protein source, but are usually short on carbohydrates (starches and sugars). In other words, we give them steak and forget to supply the potatoes or pasta. They get their vegetables in the form of plant root extrudates and organic matter or humus. The prevailing compaction eliminates needed air space which makes bacterial activity somewhat difficult to re-establish. One approach is to simply add bacteria directly to the soil and hope enough will survive to start a change. The most economical method, of course, is to add the bacteria found in manures, preferably well-composted manures.

Microbial products can also be purchased. Several types of bacterial or biological products may be needed or used to accomplish a total recovery. A basic product could contain fresh strains of aerobic, crop residue-digesting bacteria. From there, more specific bacterial or enzymatic combinations can be applied. Certain bacteria are capable of living in and digesting rather high levels of herbicides. Green and blue-green algae are available that can produce polysaccharides (long chain soil sugars) which are a component of humus. These will help flocculate (loosen) soil as well as contribute to nitrogen fixation. Special enzymes produced by selected bacteria can convert sucrose (table sugar) into polysaccharides as well. Increasing polysaccharides is an important consideration since they are the glue that binds sand, aggregates clay, and provides the biologically active carbon necessary for moisture retention. Due to increasingly erratic weather and the resultant plant stress, the polysaccharide concept alone may be reason enough to begin a biological program. A reduction in water costs can be a financially beneficial side effect of biological farming.

A Wisconsin farmer told the authors that he checks his fields regularly for signs of bacterial activity and adds bacteria both to the soil and to the plant in the form of a foliar spray. "I'm amazed at how little microorganism action existed before I started a program, even in my alfalfa crop. I couldn't find a single alfalfa root nodule in any field."

Manures and Green Manures

Providing animal manures or green manures can speed your biological recovery process. This advice must be tempered with many considerations for both. The current content of animal manures and the way they are stored or handled can be a problem. Liquid manure from a dairy that flushes their chlorine wash water into the pit tends to be putrefied rather than decomposed. How does it smell? Hog manure tends to be salty due to the amount of salt fed. Most animal manures tend to contain undigested protein and antibiotic residues which add to the problem. Proper decomposition in a liquid tank is essential as in a dry compost pile. Adding molasses, yeast, hydrogen peroxide and bacteria to either situation can help decomposition.

With or without animal manures, always consider a green manure. Green manure could consist of rye, oats (which are excellent for releasing phosphate for the following year), vetch, clover, etc., or even a green algae mentioned above. The green manure crop will help reduce excess soluble potash levels and, if sprayed with a bacterial product prior to disking into the ground, is an excellent way to bring back beneficial soil bacteria. Cover crops can be started at any stage of the economic crop growth. You don't have to wait until harvest.

Rock Powders and Humates

Dr. John Hamaker has postulated that adding such mineral deposits is directly correlated with renewed growth of plants and microbial life as demonstrated in the Black Forest of Germany. If they are powdery enough, they are readily acted upon by soil bacteria which brings them into biological availability for plant life.

Other deposits are as close as the nearest gravel pit and may be available simply for the cost of hauling. By measuring the paramagnetic qualities of the materials you apply, you can purchase only those inputs which will contribute to reducing or eliminating the need for both fertilizer and rescue chemical inputs.

Chapter 23
Levels of Non-Toxic Farming

Justifying the need to make a transition toward biological, non-toxic farming and off toxic chemicals should not even be necessary. It appears so obvious in terms of environmental and moral considerations. A wealth of scientific data showing the damage done by toxic chemicals as well as potentially awesome liability situations should tweak the interest of even the most closed-minded person. Let's start with the current products which we'll call acid, salt-elevator fertilizers and include anhydrous ammonia and dolomite lime in the same category. This is the program you should be trying to move from because these products are the source of subsequent problems which lead to the reliance on toxic chemicals needed to rescue soil, plants, animals, and people. Stop using 0-0-60 (KCl), 0-46-0 (triple superphosphate) or any blended fertilizer using the two; discontinue anhydrous ammonia and dolomite lime except in rare situations. These products must be replaced by those which don't damage soils and eventually replaced by those which only benefit soils. The following discussion will cover seven levels of fertility concepts. You will find one or a combination of several which both feel right and which you can afford. As with everything, initial cost is often the least important variable. Making a decision to change your mind, to change your approach and management practices, is usually more important than out-of-pocket costs.

Level 1

The first level toward which you may move is what we'll call *selective elevator*. At this level, you choose conventional materials like ammonium nitrate (33-0-0), ammonium phosphate (11-52-0 or 18-46-0), ammonium sulfate (21-0-0-24S), sulfate of potash (0-0-50), and 28 or 32 percent liquid nitrogen. The amounts used will be in the same range as the acid, salt level use, 100 to 300 pounds per acre. This level would also include the standard trace minerals of sulfate derivation such as zinc sulfate. Farmers may also use a variety of products from various other programs. These could include enzymes, bacterias, carbohydrates, soil conditioners and humic acids.

This first program level is certainly a transition, but can it be sustained? Will the soil biology revive? Will the brix (sugar) come up? Will the tilth come back to the soil? You won't know until you try it. Our main concern is that this program level may keep your nitrogen requirements at a level which will exceed future legal levels of application as dictated by the nitrate leaching problem. We also consider the required nitrogen levels too high for maintaining a biologically active soil.

Level 2

Hot mix liquids will be the second level. (These are arbitrary levels and do not necessarily indicate rank or choice of one level over another.) Hot mixes offer a chance to get away from dry salt problems. Nitrogen requirements may go down some, but as with all levels, everything is relative to the types of soil, previous inputs, liming, soil testing and additional products used. These are generally used at the rate of 5 to 7 gallons per acre preplant or at planting plus 3 gallons foliar.

Farmers using hot mixes have reported such concerns as price, limited availability of grades, excessive or under liming recommendations, and the most recent one is a loss of energy (umph or response) after several years of use on the same field. Hot mixes should not be used to correct CEC concepts and their use should not lead to ignoring the base saturations and pH of your soil. Hot mix desirability over acid, salt products far exceeds any negative consideration. Transitions may involve passing through several of our identified levels as well as mixing aspects of any of them.

Level 3

The third level will be called "quasi-organic" because manufacturers blend organic materials, like colloidal phosphate and leather tankage, with non-organic materials, such as 11-52-0, urea, and sometimes even with the "no-no" KCl (0-0-60). They may also blend all organic materials, but are usually noted for pushing the higher numbers, such as 10-9-7 and 6-8-4, which usually cannot be achieved with strictly organic dry materials.

Good progress can be achieved with Level 3 materials. Complaints are usually about cost or handling. Biological activation to an optimum level may be delayed if KCl or other bio-damaging materials are used in the blends.

Level 4

The fourth level will consist of single-product mined or blended materials. This refers to dry humates, clays, coals, leonardites or other deposits that claim to stand alone as fertilizer and/or limes. These products are also used to increase soil paramagnetism. The varied composition of U.S. soils would seem to preclude a single product, but several are successful in creating soil fertility changes. However, you may need to add Ca-NPK-Mg-S-type products from Level 1 or Level 2 to compensate for fertility deficiencies.

One major farmer complaint is cost due to shipping. Farmers need to select Level 4 materials from as close a geographical source as possible.

Level 5

Fifth level programs would be truly *organic*, defined as meeting qualifications for organic certification. One such type of fertilizer consists of dry blended plant residues such as cocoa shells, peanut husks, and castor bean pumice along with mined rock minerals such as colloidal phosphate, green sand, and ground granite meal/dust. Nitrogen may come from leather tankage, cotton seed or dried chicken manure.

A similar type is a composted mixture of materials as above. Both types seem to be good soil conditioners and compete against the type four (single product) materials. They seem to suffer in the energy levels available for pop-up, but usually come through nicely at harvest. Another problem area is their inability to fine tune and manipulate crops. A second aspect of this level includes such products as liquid fish, seaweed, humates (dry or liquid humic acid products), bacteria, and a number of newer products including organic trace minerals.

Level 6

The sixth level is dry soluble fertilizers. This level should automatically include the crossover to other levels since it may require lime, colloidal phosphate, selected elevator fertilizers, manure, and local rock mineral residues. Dry solubles provide the energy kick to make everything else work. Some dry soluble companies may also use a wide variety of additional products, as mentioned earlier, to attack and correct specific problems. Dry solubles themselves usually consist of clean, high-energy sources of nitrogen, potassium, potash, calcium, manganese and sulfate sulfur. They may also contain trace minerals or other additives. They are dissolved in water and used at low (5-25 pounds per acre) rates. Problems reported include lack of solubility in some water and fall out of over-saturated solutions. If used with technically accurate programs, it may require mixing of several grades to achieve grade exactness.

Having multiple grades available is a strong point for dry solubles. Lower cost for transportation and lack of contaminants are also positive points. Using dry soluble fertilizers usually call for "spoon feeding" the crop. Additional trips across the field can be avoided with proper equipment and management.

Level 7

The seventh level is biodynamics. With this program, you don't usually transport much of anything onto the farm. It works best in closed systems using livestock and home-grown feeds. It should be a spiritual endeavor akin to the proper growing of ginseng as practiced in many Oriental societies.

The main task is to use ruminant manures to capture the energies or "information" of the cosmos. This is accomplished by placing the manure in cow horns and burying them over winter. Composting and energy absorption take place while buried. Spring retrieval is followed by precise mixing into water and precise application onto the land. Other brews may be started in other containers besides horns for different agronomic purposes.

Biodynamics is used on large acreage in Australia, but currently has limited appeal in the United States. It might spread more rapidly now that biological/sustainable agriculture is expanding. The use of homeopathic approaches would also fall under this category.

Level 8

The eighth level is yet to be defined. It may involve materials or processes that gather cosmic energies (information) without the burying concept. It may involve current technology concerning cosmic pipes, radionics, or sound generation. It may also involve liquid crystals. It probably will involve specific bacteria species for root colonization on leaf surfaces. We hope to be a part of developing this level. Space programs may depend on them for long trips.

Each of these levels has its own strengths and weaknesses. The further up you go, the more attention must be paid to management. Trying to farm biologically or non-toxically while practicing standard or conventional managerial approaches is only to ask for disaster.

If you are not already in a transition, you will want to consider at what level you're going to start. The important thing is to start. Further delays could leave you on the outside looking in.

We will group our remarks into three categories: spring, summer and fall as we try to generalize the fertilizer approach in order that you can adapt the principals to your particular situation.

Additionally, we make the assumption that soil mineralization is best applied in the fall. For excellent biological activity and to take advantage of the energy of whatever you are applying, soil nutrient levels must be balanced or activated as discussed earlier. This includes Ca, P, K, Mg and S as well as accounting for the presence or lack of nitrogen.

Chapter 24
Suggested Fertility Programming

Spring Programming — Preplant

If a soil flush is needed and has not been applied in the fall, very early spring would also be an appropriate time.

As soon as primary tillage is finished or the soil is warm enough, soil and bacterial stimulation can begin. Having your equipment rigged to be able to apply a broadcast solution while doing secondary tillage would be very beneficial. The preplant application is timed to prompt soil bacteria into life. The amendments will *feed* these microbes which will then provide the new seedling with readily available plant food. This food will come from the consumption of said amendments, organic matter and humus that have already been reduced to biologically available food with a much greater affinity and efficiency for plants than simple fertilizer.

This is an excellent time to consider applying materials such as liquid fish, seaweed, biological preparations, sugar, molasses, compost, liquid calcium or any other appropriate material that will stimulate bacterial activity and raise energy levels in preparation for planting. When applying raw fish emulsion to the soil, consider a range of between 1 to 3 gallons per acre. You would be best advised to mix a good soil bacteria into the fish and let it set 6 or 7 hours prior to application. This will assist in releasing amino acids from fish protein. This step will not be necessary when using the new fish hydrolasates, since their production includes enzymatic breakdown.

Regardless of whether or not the above steps are used, it may be advisable to consider a low to moderate preplant nitrogen program about two weeks prior to planting. This is to provide electrolyte to the soil "battery," so it is in a charged condition when the seed goes in the ground. Soil temperature is an extremely important consideration at this time. Soils which are wet and cold will often germinate poorly. It is best to delay planting if possible as most crops will mature more rapidly when planted later, once the soil is working as we've described. If it is necessary to force the issue, using a strong growth fertilizer may help overcome the lack of growth energies available due to the low temperature. Products such as liquid calcium, or calcium nitrate may provide those energies.

Using acid fertilizers like 0-46-0 gives a tremendous energy push to the soil during times of cold temperature which helps account for conventional farming success even though planted early. This same heat or activity, however, is what creates the formulation of tricalcium phosphate in which neither the calcium nor the phosphate are available to the growing plant. Preplant fertilizer is best applied broadcast or sprayed. This approach will cover the soil, encouraging bacterial response over the entire soil surface.

Planting

When planting occurs, many farmers feel in-row support to be critical. The authors feel that, although in-row support may be important, it can often be achieved with as little as 5 to 10 pounds of dry soluble fertilizer, a few gallons of a hot mix, and a few gallons of 10-34-0 or 200 pounds of an organic mix.

Energy Release vs. Crop Need

In the conventional program, dry fertilizer is applied in typically large amounts at planting with no fertilizer applied again until next year's crop is planted or possibly a sidedress. In sustainable programs, fertilizer may be applied several times throughout the year: preplant, planting, foliar and/or side dressing at fruit filling, and dry down. In the conventional program, the fertilizer energy is mostly lost to the crop while in the sustainable program, energy is managed throughout the growing year for the benefit of the crop.

Conventional programs provide an extreme excess of energy to seedlings which only require a very small amount. Once the crop becomes firmly established, however, and reaches its maximum energy requirements, conventional fertility programs are usually past their peak and are unable to supply the potential useable energy needs of the crop. This results in one ear of corn instead of the several that can emerge, one on each node.

Conventional fertility programs apply massive amounts (a year's supply) of fertilizer at planting while the seed and/or seedling need relatively small amounts of energy. This energy expends itself in heat and other ways and the energy gradually dissipates. The heat released helps account for the fact that seeds survive early planting when the soil temperatures are too low. Conventional programs can work when soils are cold, whereas biological or non-toxic programs may fail due to the lack of heat released. Later in the season the energy levels are nearly expended at the same time the plant is now requiring them. To counter this effect the suggested programs make use not only of biological activity which continually releases energy, but also of side dressing and foliar feeding. Both applications provide the plant and soil with energy inputs. This is the key to low-fertility input programming. Energy release is essential to the proper growth of plants including seed formation, seed fill and dry down. When the energy is managed properly, the system works.

Row Alignment

Because the soil attracts a magnetic flow traveling from the south to the north pole, the farmer wants to arrange his planting so as to maximize the energy obtained from that magnetic flow. For this reason, Dr. Reams recommended planting row crops in an east-west direction rather than a north-south direction. When planting in an east-west direction, the roots will tend to grow toward the north and grow between the rows. The same plants would grow into the root mass of the plant to the north of it when planting in a north-south direction. This alignment would cause a competition for nutrients. Beds for strawberries may do better on north-south layouts. Perennial crops, especially those grown on metal trellis structures should be planted on a north-south alignment so that the plant can take advantage of the magnetic attraction to the metal structure.

Summer Programming

Summer programs consist basically of foliar sprays and side-dressing. Foliar sprays should be nutritional in nature, with rescue chemicals added only if absolutely needed. Foliar sprays are used to enhance growth, to influence seed formation during the developmental phase, to push a plant into a fruiting response, to influence the size of the developing seed and, of course, to apply needed nutrients and trace minerals. Research from Russia shows that increased nutrient uptake occurs in the root zone where beneficial bacterial response is noted following a foliar spray. Side dressing is a way of providing an additional energy boost as the crop requires it. This approach will benefit soil bacterial activity as well. Some nutrients, like calcium, do not travel easily within the plant; therefore, foliar sprays can be an excellent method of application to assure availability in the plant.

Crops do not realize genetic potential due to many factors including lack of water at critical times, lack of a balanced nutrient supply, lack of soil mineralization, and lack of energy when needed. Water needs are especially critical. The closer the LaMotte test readings match Dr. Reams' recommendations, the more critical water availability becomes. As humus levels increase, the plant is much better able to meet its water needs even in dryer conditions. Walking a field in early morning, you can find almost twice the dew on a biological field compared with a conventional neighbor. This additional water makes the difference between the two crops.

Mid-summer is a critical time for most crops because as fertilizer energy is being depleted plant needs for energy are increasing. Foliar feeding is one energy source, sidedressing is another. By sidedressing a fertilizer designed to assist the crop in producing fruit and in stimulating the release of soil energy, you provide the plant with energy needed to begin to live up to its genetic potential.

Equipment Setup

On a very practical note, how do you accomplish sidedressing and foliar spraying without spending too much time or money? The solution is to set up your equipment to handle multiple steps in a variety of ways.

Modify your tractor and equipment to meet the needs of today's timely, spoon-feeding techniques. Two saddle tanks are a necessity whether homemade or purchased. You can get away with only a field sprayer for just so long, and then the trips across the field become too costly. Let's see what you could do with saddle tanks.

The materials in the saddle tanks have to be applied to something via some outlet. Use a front spray boom that can foliar feed and/or drop-nozzle materials onto the ground. A second tank, or adapter off a single tank, should be able to direct materials through a squeeze pump into cultivators or planters. For the final step, place a boom adapter behind the drill or planter. Now you can do the following:

1. Apply preplant material behind a primary tool (chisel plow).

2. Apply preplant material from the front boom while doing secondary tillage and work it in.

3. Apply two incompatible materials preplant from two tanks with front and rear booms while secondary tilling.

4. Apply two mixtures while pulling a field sprayer with a third mixture.

5. Apply one or two mixtures at planting, one in the row and one sprayed behind the planter for weed control (calcium and molasses), or one in the row and one broadcast fertility program off the front boom.

6. Apply a foliar at 2 to 6 inch stage while sidedressing a different mixture with the cultivator.

7. Sidedress soil materials at last cultivation (conversion stage) while foliar spraying a different plant manipulating mixture.

8. Crown feed hay or alfalfa during cutting by setting up angle jets underneath your haybine. Cut hay will land on part of the sprayed spot, but that will not affect feeding if crop is baled or chopped during the usual time frame.

9. Use your imagination!

This solution is intended to address the common complaint concerning spoon-feeding or crop manipulation of too many costly trips across the field. You may be able to complete a simple program in three trips for row crops: preplant, plant, and last cultivation. Feeding at 2 to 6 inches won't add a trip if you are cultivating anyway.

Berries, Fruit and Nut Trees

Orchards, including berries, are programmed the opposite of field crops. In field crops, the seed must sprout and grow before the plant is ready to reproduce. In trees and berries, the fruiting process usually begins as soon as the plant breaks dormancy in the spring. Trees and berries will greatly benefit from foliar feeding. They could even benefit from foliar sprays during the late winter months when no leaves are present, as proven by Michigan State University and the Atomic Energy Commission in the 1940s. Materials are often applied during this time to prevent wind damage and drying out. In early spring, begin with a fruiting or reproductive spray as early as green tip. This can continue until the fruit is well set on the tree. When the tree is in bloom, it is best to refrain from spraying during the pollination period.

Foliar sprays should be purchased or formulated so that phosphate is contained in most applications along with a small amount of nitrogen to act as an electrolyte. At this stage, the nitrogen used should be confined to urea or ammonia nitrogens or from natural, low-dose sources such as fish emulsion. This is to avoid the pushing of upper tree growth at this time. If the upper tree is growing, the new roots are not growing as well as they should and the hormonal output of the tree will be encouraging growth rather than reproduction. Materials such as fish and seaweed should be considered standard as well. Additional materials could include calcium, potash, vitamins, trace minerals, hydrogen peroxide, garlic sprays, spices, bacteria, compost tea, etc.

As mid-summer approaches, it is time to switch to growth sprays on perennials. These would usually be based on a non-nitrogen form of calcium, preferably in a chelated form involving a sugar or plant acid, and potassium in a carbonate or hydroxide form. Switch to potassium in a thiosulphate form when you're ready to "finish off" a crop. Sprays should be continued until dormancy, but discontinued about two weeks prior to and during harvest unless fruit is low brix and needs rescuing.

For better storage, try using a chelated calcium with garlic, which is highly systemic, just a week or two before harvest. Continued feeding after harvest may be advisable if the harvested crop was exceptionally heavy and ground fertility levels are lacking.

Fall Programming

Mineralizing

Mineralizing usually costs the most money in the short term, so it gets ignored too often. This is why we suggest locating and using free or inexpensive, nearby natural minerals where possible. Lime or marl are part of the mineralization process and usually have to be purchased, but gravel or kiln dust may be available for the hauling. Basic slag from industry is an underused possibility. Relatively few livestock farms will require potash mineral supplementation because of the large manure applications applied. For those of you who truly need potash, sulfate of potash (0-0-50) is available from major elevators. Some organic certifiers are now approving 0-0-50. For a broader, mineral-based potash, try greensand at 7 percent or granite dust at 5 percent. Sul-po-mag, a complex molecule containing both magnesium and potash sulfates, is another possibility. However, if you apply it other than between July 15 and September 15, it will break down into its two components rather than perform its special effect as a copper release agent and aid in preventing bark or fruit splitting in rapid growth situations (see Chapter 11).

Most of the above-mentioned mineral sources will contain a variety of trace minerals so you don't have to seek out a special dry, bulk additive of a particular trace mineral unless soil tests indicate otherwise. If you get a blend that contains humates or compost, you are also getting trace minerals.

Humus

Building humus or active carbon levels are also important fall functions. Everything you've done to create a biological system will naturally increase the conversion of existing raw organic matter into active humus. Humus is amorphous (without specific form) and has no real formula. It contains several factions of acids, such as humic, fulvic, and ulmic, as well as active carbon sources such as polysaccharides (soil sugar/glue). You need to be interested in humus, but also recognize that it is hard to distinguish between digested plant residue (humus) and dead bacterial bodies (protoplasm) which actually combine various components to form stable humus.

Most of you know that you build humus by plowing down manure or crop residue and allowing the soil to compost. You are usually told that it is impossible to raise humus levels very rapidly. Many farms are doing it rapidly by using biological methods which stimulate microbial activity. These include application of humates, coals, sugars, good fertilizers, additional bacteria, detoxifying substances, and enzymes. Molasses works well because it not only provides energy for microbes, it also supplies trace minerals, such as iron and sulfur. Sugar has played a major part in our active carbon-building programs because farmers can convert sucrose or table sugar directly into 10,000 to 15,000 carbon chain polysaccharides in a matter of a few days or weeks by biological activity.

General *Dos* and *Don'ts*

Some suggestions for avoiding problems, especially in blending or applying new products, may be helpful. Also included are suggestions that can give a biological program an even break in the transition period. All the following suggestions and cautions are based on growers' successes, mistakes and observations.

A. Mixing Guidelines and Cautions:

    1. Incompatibility: The following materials may cause problems if mixed together:

       a. Liquid calcium and humic acid.

       b. Liquid calcium products with phosphorus or sulfur-containing products.

       c. Biological (bacteria) products with any strong nitrogen, sulfur or toxic materials.

       d. Liquid calcium products with hard water.

    2. Products difficult to mix or spray:

       a. Wood or beet molasses (may have fibers).

       b. Biologicals (bacteria) grown on bran or other insoluble food sources.

       c. Dry soluble fertilizer containing calcium mixed in hard water.

B. General Don'ts:

    1. Don't mix dry soluble fertilizers intended for 20 gallons of water in 5-gallon pails and transfer to the tanks. Use as much water as possible for initial mixing.

2. Don't dump dry soluble fertilizer into spray tanks and suck them directly into the pump. Pour in slowly or use a short baffle pipe to draw water off at mid-tank level.

3. Don't get colloidal phosphate wet before spreading.

4. Don't add hydrogen peroxide to concentrated mixtures. Always add hydrogen peroxide to the water first.

5. Always pour acid into water, not vice versa.

6. Don't work carelessly around concentrated acids such as phosphoric or sulfuric. These can burn or blind.

7. Don't spray fertilizers on tender foliage in the heat of the day.

8. Don't work unprotected around toxic materials. The crotch (and scrotum) are highly susceptible to absorption. Also, watch where your hands go during the day and especially at night!

C. Management Guidelines

1. Don't plow or work ground while it is wet. The compaction effects can last two to three years.

2. Don't overwork ground.

3. Don't apply large amounts (more than 1 ton) of lime in the spring.

4. Don't over apply ammonium sulfate (more than 200 pounds), especially in the spring on annuals.

5. Don't plant in cold, wet ground. Wait for 60 degree soil temperatures at seed depth.

6. Don't plant high populations if you're seed treating and fertilizing with good materials. Higher than normal stands may reduce yield.

7. Cultivate at least once for weed control and to allow air into the soil.

8. Foliar feed at early stages (4 to 6 inches) to affect yield, and at later stages to convert fertilizers from growth to fruiting or from fruiting to growth (trees).

9. Sidedress and foliar feed for fruit fill.

10. Always get a cover crop in somewhere, somehow each season.

11. Monitor soils and crops involving high-value annuals each week at a minimum with a refractometer and soil instruments and react to problems immediately.

12. Watch for nitrogen deficiencies on high-magnesium soils.

13. Think positively. Mental input sets the stage for actual outcome.

Finding a Consultant

Find and use the services of a biologically oriented consultant. In some sections of the country, this may be difficult. Read magazines, newspapers, or attend conferences to find out who is working the area.

Consultants need to be paid for their work. Their pay comes either in the product they sell or in the services they offer. Sometimes more objective information can be obtained if no products are sold, but sometimes not. Consultants who are proficient at what they are doing often choose to market products because the standard outlets do not have the right materials available. They become familiar with a line of products that have proven effective on the crops in their location and they can ensure that the products will be made available to farmers in the area. Give your consultant a chance for you to see results. One year is not a good trial although many changes can occur in this time frame. If you expect to see results in one year's time, say so up front. It is preferable to give a program a minimum of three years to perform. It has taken many years to create the conditions which now exist, and they won't be overcome quickly in most cases.

Take time to choose a consultant. Ask neighbors, talk to satisfied customers, and call around. Don't expect glossy advertisements, but do expect to speak with someone who sounds knowledgeable. Realize that these consultants are finding themselves increasingly in demand. A few years ago the work and goals they choose were generally discounted by the "system." Now there aren't enough to go around, especially those who are trained/experienced and who are effective in their work.

Costs?

Anyone should be able to take a plot of pure infertile beach sand and make it into living soil for under $200 per acre. This could include lime, phosphate, fertilizer, manure and rock mineral powders. It usually takes a lot less money to recover reasonable farm soil. It makes more financial sense to spend recovery money on your current fields than to purchase new ground for $750 to $2,000 an acre and pay taxes on it every year. *Farm deeper, it's cheaper.*

Finally, don't quit! Non-toxic, biological, alternative, low-input, sustainable (or whatever you prefer to call this type of farming) agriculture is not a passing fad. It is a long overdue correction to a simplistic, limited approach to solving farming challenges.

Chapter 25

Subtle Energies

By now, readers may be wondering why the information in this book differs so greatly from what is understood as conventional agriculture. You might also be asking if what you have read is true. Truth in the scientific world is a fleeting thing as each new discovery either validates or invalidates that which was previously thought to be true. Therefore, one has to discern truths about individual concepts and see if those individual concepts add up to an observable, acceptable whole.

One doesn't have to look at modern agriculture very long or hard to see that the individual concepts concerning the necessity or results of using NPK salt acid fertilizer does not add up to what is acceptable. The litany of polluted water, eroded land, low brix crops, antibiotic-dependent livestock, depressed rural economics, poor health statistics, Mad Cow disease, antibiotic-resistant human attacking germs, *E. coli* contamination of meat and cider, breakdown of the family, crime and drugs (yes, they are all related), is not exactly what we would say is acceptable.

In other words, the science applied to agronomy was not scientific in the true sense. It was pseudo-science masquerading as science because it blindly followed reductionist science procedures without observing the whole picture or relating the results to the reality of what nature is or how it works. It was also pseudo-science because it chose artificial end results and directed the experiments to prove those results. Modern agronomy applied linear (straight line correlations) thinking and linear procedures to a non-linear (complex) system called *life* and got the wrong answers. Thomas Kuhn, author of *The Structure of Scientific Revolution*, showed that scientists are capable of completely shutting out or ignoring experimental data that doesn't agree with their preconceived notions. In other words, they unconsciously "fudge."

Luckily, the pressure of oil and chemical money on institutions has slowed enough to stifle much of that kind of research. Environmental pressures are now empowering holistic (non-linear) thought processes and research as well as reductionist research which doesn't start out with preconceived notions.

The Nature of Reality

Most adults today were educated under Newtonian concepts. Albert Einstein and other scientists proved that the concepts of quantum physics and quantum mechanics were much better at explaining what goes on at the sub-atomic level. Discoveries in quantum physics show that, in some cases, experimenters cannot find something until they look for it because their consciousness is necessary to make it appear. Light, for instance, travels in straight lines but also bends. Light particles can go both forward and backward in time. Michael Talbot has written a fascinating book, *The Holographic Universe*, describing these findings in layman's language.

The discoveries and understandings founded in quantum physics give a new meaning to many aspects of life formerly not understood. How can people *know* something which hasn't taken place yet? How was the Hebrew prophet Samuel able to tell shepherds where their animals had run off to even though he was nowhere in the vicinity? How were people able to pierce their flesh with nails, often under religious conviction, and neither bleed nor suffer cuts or bruises? How are people able to cure themselves of such dreaded diseases as cancer merely via the use of their minds? In a similar vein, Robert B. Stone reports on Cleve Backster's latest research in *The Secret Life Of Your Cells*. Backster showed that plants have a consciousness and if you monitored them with equipment such as is used to record EKG or EEG readings in humans, you would find that the plant would respond to circumstances of life. For instance, Backster placed several plants in a room and had someone enter and utterly destroy one of the plants. Subsequently he then had several people enter the room. When the person who destroyed the plant entered the room the plant registered readings indicating that he was the "one."

The Holographic Universe and *The Secret Life of Your Cells* help provide an understanding needed to comprehend this handbook. All of life has a consciousness at one level or another. When you curse your soil, animals, equipment, employees, children, wife, etc., you send out energy which is counterproductive to healthful living. Conversely, when you bless your soil, animals, equipment, employees, children, wife, etc., you send out energy which builds up, empowers and gives life.

Elusive Energies

In the field of human health Dr. Samuel Hahnemann, in the 1800s, discovered what he called homeopathic medicines. These consisted of mineral or herbal substances repeatedly diluted to the point where no actual substance remained. They were also shaken after each dilution. This process forced the energy of the original substance in between the water molecules leaving nothing but an *energy* or *vibration* in solution. Homeopathic remedies are used throughout Europe today as an alternative to conventional allopathic medicine as practiced in the United States. We believe the homeopathic approach is much closer to the reality of how life functions than our current allopathic approaches.

Dr. Callahan has proven that most of the processes of life operate as non-linear interactions of electromagnetic waves in the infrared, visible and ultraviolet spectrum (see *Exploring the Spectrum*). In simple terms, you can't add signal A to signal B and get signal A + B in living systems. You add A to B, and you get C, which is a whole new, not necessarily predictable, signal.

The patenting of radionics, circa 1950, and the invention of the magnetic nuclear resonating medical systems in the 1970s seem to prove the existence of additional subtle energy emissions from matter. The detection of brain waves and biofeedback further strengthen the ideas of subtle energies being able to affect life.

In Europe, photomicrographs of energy emissions from below the earth's surface have been correlated with disruption of the human body's own electrical system to a point of disease formation and death. Dr. Fritz Albert Popp of Germany has proven that each living cell is a tiny capacitor with a coherent charge (cloud) of electromagnetic energy superimposed on the physical material. Loss of coherence of that cloud is loss of immunity and the complete loss of the cloud (measured as mitogenic radiation) is death.

The above findings can be translated into useful, informative procedures for any farmer who wishes to use them. The procedure involves using the human mind and body to interact with the subtle energies of nature in order to perceive what our normal sense doesn't ordinarily recognize. The animal kingdom has extensive examples of its ability to detect and react to nature's subtle energies. Examples of this phenomena include migrating birds and mammals, feeding sharks, domestic livestock's reaction to low-level, stray voltage, and many animal reactivities to impending earthquakes. There is no reason for man not to take advantage of the same natural gifts. However, few of us are gifted enough to be able to consciously sense as low an intensity of energy as an animal. To overcome our conscious limitations, man can resort to kinesiology or dowsing which are non-conscious reactions.

Kinesiology

Many health practitioners and chiropractors practice a form of muscle testing known as kinesiology. The practitioner places his hand on the outstretched arm of another and presses down while measuring the amount of strength it takes to press the arm down. The person holding his arm out resists the efforts to push his arm down. Together they both assess how much strength it took to move, or not move, the arm. Next the person is given a substance to hold, usually below his heart. This can be a sugar cube, a sample of fertilizer like 0-0-60 (which is also sold as table salt substitute), an apple, cigarettes, etc. The person again resists having his arm pushed down. Again, both persons assess how much strength was needed, preferably with the person doing the pushing (testing) keeping his efforts the same in both situations.

The results will fall into one of three categories: the person tested will test the same, stronger, or weaker. If he tested weaker, the substance weakened his vitality. If stronger, it strengthened his vitality. If he tested the same, his vitality remained relatively unchanged. By testing several items, the person quickly comes to realize that different substances, foods, fabrics (such as wool, cotton, or synthetic fabrics), all have a factor in strengthening or weakening his vitality. Now test with simply a thought in mind, for example, think of a time when you felt especially peaceful in your life. Test. Now think of a time when your life was particularly troubled and you felt your world was caving in on you. Test. Think of some beautiful, lovely music. Test. Think of some of the unpleasant music being played today. Test.

You can see the possibilities from these sample questions. You will generally find that people will test strong to foods that are good for them, and weak to those toward which they have an allergy. People will test strong to natural fabrics and weak to synthetic fabrics. When testing a salt substitute (0-0-60), most people test weak, the same as their soils. Although kinesiology would not be considered scientifically accurate by the American Medical Association, it can be used for good approximations. Several farmers use themselves as an intermediary to check feed inputs for livestock. Put the substance to be checked in the right hand and place the substance below the heart. Have a second party check the relative strength of the extended left arm before and after.

The scientific explanation involves the infrared signals coming from any and all substances which either resonate with and strengthen it, or create dissonance and weaken it. The life form subjected to those specific emissions is going to react negatively, neutrally or positively.

Magnetism and Ley Lines

Because of the subtle nature of magnetism, most conventional biological or agronomic science has paid little attention to it as a subject of interest. The ancient Egyptians apparently were aware of paramagnetism and diamagnetism as they had different symbols for each. The Chinese also have a long history of measuring and considering magnetics in terms of the placement of villages and homes. In modern times, we have seen the use of magnets on water lines for conditioning, for stimulating healing on living tissue, as a replacement for X-rays in the nuclear magnetic resonance medical machines, and as a factor in space travel studies.

The magnetic field of the Earth is all pervasive and, therefore, affects all soil, plants and livestock. The magnetic field is not evenly distributed over the Earth as assumed. Grid patterns of higher intensity areas exist in a basic north-south and east-west direction. Several levels of interlocking grids actually exist, but the one important to local agriculture has a distance between the stronger lines of approximately 20 feet. The British term for these anomalies is "ley lines." Ley lines have significance because they can be disruptive to those little electrical capacities called living cells. Many farmers have discovered that cows standing in certain stanchions will always have health problems and may try to avoid entering the stanchion. Identifying these locations as ley line intersections and interfering with the grid signal has alleviated the problem. Farmers have also observed poor feed storage in particular bins that have turned out to be at a grid intersection.

Dowsing

One can pay several hundred dollars for a magnometer to detect and measure variations in the magnetic field, or one can bend two 25-cent coat hangers (two $1.00 brass welding rods are better) and have a subtle energy detector to connect to the rudimentary lodestone deposit area in your brain. (A homing pigeon has a clearly-defined lodestone mineral deposit.) Bend the rods in an "L" shape. Grasp the shorter end of the L rod, point the long end away parallel to the ground, leave space between the rod and your forefingers for the rod to swing left or right, and you are ready to find ley lines by dowsing.

Walk slowly in a north-south or east-west direction. Activate the left side of your brain by speaking out loud, "When I encounter a ley line, the rods will separate and point parallel to the ley line." (Sometimes they cross instead). Activate your right brain by thinking the same thing as you're saying it. Approximately seven out of ten adults will detect the lines, but 95 percent or more children will usually be able to detect ley lines. Once you've mastered ley lines, then you can try water lines, underground electrical wires, water veins for well drilling, etc.

Further advancement can be made by reducing your dowsing tool to a small object on a string or chain. (This allows you to do daylight dowsing so the neighbors won't see you running around the yard with two welding rods while talking aloud and looking like a complete idiot.) Dangle the object (a metal nut, a macrame ball, etc.) over your fingers on a 3- to 6-inch cord. Train your mind to cause the object to swing back and forth (away and toward you) when a yes answer exists. (Obviously you can only obtain answers to yes or no answers). Train your mind to cause a sideways motion to occur for no answers. Now these readings can be used to have your livestock, crops, plants, trees, etc. indicate to you whether a particular input is good, neutral or negative to them. A sample of an input (fertilizer, sugar, sprays, etc.) is placed next to or near a crop, plant or livestock. By phrasing the question such as, "Will this fertilizer raise the vitality of this tomato plant," your mind can interpret the interactions of the emissions of the product and your plant. The dowsing object will move back and forth in one direction or the other to indicate a positive or negative response. Judgements can be made as to the appropriateness of the product to the plant.

Because of the economics of commercial production, most growers would prefer to use a more sophisticated method of testing bioelectric interactions such as an electronic scanner. However, home gardeners and commercial growers may find the dowsing method handy for many situations.

Persons finding the subject of dowsing unfamiliar to them or perhaps doubting its authenticity should refer to the American Society of Dowsers for more information. They have extensive files on national and international scientists and engineers with information and research on the subject. A marvelous book entitled *The Divining Hand*, by Christopher Bird covers the subject of dowsing like no other.

We trust you have benefited from the information we've provided here. Use that information which is appropriate for you, discard the remainder. If you discover flaws in our model and/or have mastered more of the physics, chemistry, biology and spiritual nature of the universe than we, write an update on this important topic. God bless.

Resources

The field of non-toxic farming is literally exploding. Where 20 years ago you could count suppliers on your fingers, now there are hundreds. Following are contact addresses and phone numbers for a few of the companies mentioned in the text of this book. For additional suppliers refer to any issue of *Acres U.S.A.* and the various state and regional organic association newsletters.

Crop Services International, Inc., 1718 Madison S.E., Grand Rapids, Michigan 49507, toll-free 1-800-260-7933, phone (616) 246-7933, fax (616) 246-6039.
Provides soil and plant tissue testing, rock dust paramagnetism testing, fertility recommendations and on-site consulting drawing upon all of the disciplines outlined in this book. Owned and operated by Philip and Louisa Wheeler.

Acres U.S.A., P.O. Box 91299, Austin, Texas 78709, toll-free 1-800-355-5313, phone (512) 892-4400, fax (512) 892-4448, e-mail <info@acresusa.com>, website <www.acresusa.com>.
Monthly magazine of commercial-scale ecological agriculture. Now well into its second quarter-century of publication, each issue

profiles successful non-toxic farmers, covers news and developments in sustainable agriculture, and carries feature articles on all facets of chemical-free growing. All major suppliers of non-toxic inputs advertise in this publication. Acres U.S.A. also operates a mail-order bookstore carrying hundreds of titles from all over the world on ecological agriculture.

Pike Agri-Lab Supplies, Inc., P.O. Box 67, Jay, Maine 04239, phone (207) 897-9267, fax (207) 897-9268, e-mail <info@pike agri.com>, website <www.pikeagri.com>.
Manufacturer and mail-order supplier of test equipment for farmers including all meters and devices needed for Reams-style agricultural analysis, the Callahan paramagnetism meter, LaMotte test kits, and more.

American Society of Dowsers, Inc., 99 Railroad St., St. Johnsbury, Vermont 05819, phone (802) 748-8565.
Group formed to promote the use and knowledge of dowsing. Presents workshops, a large conference, and supplies educational materials on dowsing.

Agri-Dynamics, P.O. Box 267, Martins Creek, Pennsylvania 18063, phone (610) 250-9280, fax (610) 250-0935, website <www.agri-dynamics.com>.
Producer of alternative remedies and nutraceuticals for livestock and pets.

Bibliography

Many of these books may be purchased from *Acres U.S.A.*, P.O. Box 91299, Austin, Texas 78709. Check your library for the others.

Albrecht, William A., *The Albrecht Papers, Volumes I and II*. Charles Walters, Jr., ed. Acres, U.S.A., Austin, TX. 1975.

Andersen, Arden B. *The Anatomy of Life & Energy in Agriculture*. Acres, U.S.A., Austin, TX. 1989.

Andersen, Arden B. *Science in Agriculture*. Acres U.S.A., Austin, TX. 1992.

Ankerman, Don and Richard Large, Ph.D. *Soil and Plant Analysis*. A&L Agricultural Laboratories, Inc., Fort Wayne, IN.

Aspelin, Arnold. "Pesticide Industry Sales and Usage, 1994 and 1995 Market Estimates." U.S. EPA, 1997.

Baver, L.D., W.H. Gardner & W.R. Gardner. *Soil Physics*. John Wiley & Sons, New York. 1972.

Bearden, Thomas E., *Excalibur Briefing*. Strawberrry Hill Press, San Francisco, CA. 1988.

Becker, Robert O., MD. *The Body Electric*. William Morrow, New York. 1985.

Bird, Christopher. *The Divining Hand*. Whitford Press, Atglen, PA. 1993.

Bowles, Joseph E. *Physical and Geotechnical Properties of Soils*. McGraw-Hill Book Co., New York. 1979.

Callahan, Philip S. *Ancient Mysteries, Modern Visions*. Acres U.S.A., Austin, TX. 1984.

Callahan, Philip S. *Exploring the Spectrum*. Acres U.S.A., Austin, TX. 1994.

Callahan, Philip S. *Nature's Silent Music*. Acres U.S.A., Austin, TX. 1992.

Callahan, Philip S. *Paramagnetism, Rediscovering Nature's Secret Force of Growth*. Acres U.S.A., Austin, TX. 1995.

Callahan, Philip S. *The Soul of the Ghost Moth*. The Devin-Adair Company, Old Greenwich, CT. 1975.

Callahan, Philip S. *Tuning Into Nature*. Acres U.S.A., Austin, TX. 1975.

Carson, Rachel, *Silent Spring*. Houghton Mifflin Co., New York, NY. 1962.

Clifton, C.E., Ph.D. *Introduction to the Bacteria*. McGraw-Hill Book Company, Inc., New York. 1958.

Cocannouer, Joseph A. *Weeds Guardians of the Soil*. The Devin-Adair Company, Old Greenwich, CT. 1950.

Davis, Albert Roy and Rawls, Walter C. Jr., *Magnetism and Its Effect on the Living System*. Acres U.S.A., Austin, TX. 1996.

Editorial Advisory Board. *Farm Chemicals Handbook '90*. Meister Publishing Company, Willoughby, OH. 1990.

Engelken, Ralph & Rita and Patrick Slattery. *The Art of Natural Farming and Gardening*. Barrington Hall Press, Greeley, IA. 1983.

Follet, Roy H., Larry S. Murphy and Roy L. Donahue. *Fertilizers and Soil Amendments*. Prentice Hall, Inc., Englewood Cliffs, NJ. 1981.

Gershuny, Grace, and Joesph Smillie. *The Soul of Soil: A Soil-Building Guide for Master Gardeners and Farmers*. Chelsea Green, White River Junction, VT. 1995.

Jensen, Bernard and Mark Anderson. *Empty Harvest*. Avery Publishing Group. Inc. Garden City Park, NY. 1990.

Juhan, Deane. Job's Body, *A Handbook for Bodywork*. Station Hill Press, Inc., Barrytown, NY. 1987.

Kervran, C. Louis. *Biological Transmutations*. Grain and Salt Society, Arden, NC. 1988.

Keyes, Ken, Jr. *The Hundredth Monkey*. Vision Books, Coos Bay, OR. 1981.

Kinsey, Neal and Charles Walters. *Hands-On Agronomy*. Acres U.S.A., Austin, TX. 1993, 2006.

Kuhlman, Kathryn. *I Believe in Miracles*. Prentice-Hall, Inc., Englewood Cliffs, NJ. 1962.

Lakhovsky, Georges. *The Secret of Life*. Noontide Press, Newport Beach, CA. 1988.

Lisle, Harvey. *The Enlivened Rock Powders*. Acres U.S.A., Austin, TX. 1994.

McCaman, Jay. *Weeds!!! Why?* McCaman Farms, Sand Lake, MI. 1985.

McCaman, Jay. *Weeds and Why They Grow*. McCaman Farms, Sand Lake, MI. 1994.

McLeod, Edwin. *Feed the Soil*. Organic Agriculture Research Institute, Graton, CA. 1982.

National Academy Press. *Alternate Agriculture*. Natural Resource Council. Washington, D.C. 1989.

Ott, John. *Health and Light*. The Devin-Adair Company, Old Greenwich, CN. 1973.

Pfeiffer, Ehrenfried E. *Weeds and What They Tell*. Biodynamic Farming and Gardening Association. Junction City, OR. 1970.

Pimentel, David. *Techniques for Reducing Pesticide Use: Economic and Environmental Benefits*. John Wiley and Sons, Ltd., New York. 1997.

Poirot, Eugene M., *Our Margin of Life*. Acres, U.S.A., Austin, TX. 1978.

Sampson, R. Neil. *Farmland or Wasteland*. Rodale Press, Inc., Emmaus, PA. 1981.

Schriefer, Donald L. *From the Soil Up*. Acres U.S.A., Austin, TX. 1984.

Schroeder, W.L. *Soils in Construction*. John Wiley & Sons, NY. 1980.

Skow, Daniel L., D.V.M. *The Farmer Wants to Know*. 1979.

Skow, Daniel L., D.V.M. and Charles Walters. *Mainline Farming for Century 21*. Acres U.S.A., Austin, TX. 1991.

Sparrow, H.O. *Soil at Risk*. Ottawa, Canada. 1984.

Stone, Robert B. *The Secret Life Of Your Cells*. Whitford Press, PA. 1989.

Talbot, Michael. *The Holographic Universe*. Harper Perennial, NY. 1991.

The Furrow, Deere & Company, Moline, IL.

Tompkins, Peter and Christopher Bird. *Secrets of the Soil*. Earth Pulse, Anchorage, AK. 1998.

Tompkins, Peter and Christopher Bird. *The Secret Life of Plants*. Harper and Row, NY. 1989.

Traynor, Joe. *Ideas in Soil and Plant Nutrition*. Kovak Books, Bakersfield, CA. 1980.

USDA Study Team on Organic Farming. *Report and Recommendations on Organic Farming*, U.S. Government Printing Office. 1980.

Walters, Charles, Jr., *Eco-Farm — An Acres U.S.A. Primer*. Acres, U.S.A., Austin, TX. 1979.

Walters, Charles, Jr. *Weeds — Control Without Poisons*. Acres U.S.A., Austin, TX. 1991.

Weir, David and Mark Schapiro. *Circle of Poison*. Food First Books, Oakland, CA. 1981.

Willis, Harold. *The Coming Revolution in Agriculture*. A-R Editions, Inc., Madison, WI.

Index

A Handbook for Bodywork, 9
acid, 19, 210
acidosis, 149
Acres U.S.A., xiii-xiv, 8, 221
actinomycetes, 16
ADF test, 146
aerobic decomposition, 50
aerobic soil, 50
aerobic zone, 177
aflatoxins, 150
Agri-Dynamics, 154
agriculture, alternative, 212; biological, 8, 199, 212; non-toxic, 212; sustainable, 30, 199, 212, 222
Alar, 6-7, 133
Albrecht, William A., xiii, 35, 48, 59
aldehydes, 45, 68, 92, 95; embalming effects of, 92
alfalfa weevil, 29
algae, 16, 78, 192
alkali oxalates, 189
alkaline, 19
alternative agronomy, 188
American Medical Association, 217

American Society of Dowsers, 220, 222
amino acids, 16, 50, 79, 99, 131, 191, 202; formation of, 96
ammonia, 49, 79, 166
ammoniated phosphates, 82, 87
ammonium laureth sulfate, 189
ammonium nitrate, 72-73, 81, 196
ammonium nitrogen, 44, 113
ammonium phosphate, 196
ammonium sulfate, 40, 63, 72-73, 82, 97, 196; color of 82
ammonium thiosulfate, 63
amylase, 159, 162
anaerobic pits, 158
anaerobic secretions and pH, 105
Anatomy of Life and Energy in Agriculture, The, 134
Andersen, Arden B., 8, 9, 134
anhydrous ammonia, 2, 17, 40-41, 79, 81, 94, 113, 182, 195; amino acid compounds in, 81; avoidance of, 80, 87; handling of, 80
anhydrous knife, 174
antibiotic residues, 193

antibiotics, 16, 84, 139, 149
aphids, 28
apple cider vinegar, 125
aqua ammonia, 80, 81
arsenic, 98; and human heart, 132
Aspergillus oryzae, 163
Association of American Plant Food
  Control Officials, 98
Atomic Energy Commission, 123
atrazine, 140
Backster, Cleve, 215
bacteria, 16, 18, 37, 45, 68, 77, 83,
  99, 171, 187, 192-193, 196, 198,
  201-202, 208-209; aerobic, 22, 69-
  70, 192, 273; anaerobic, 68, 173,
  179; foliar form, 193; killing of, 90;
  Rhizobia, 78; specific species of,
  199
bacterial action, 69
bacterial activity, 41, 192-193
bacterial response, 202
bacterial stimulation, 81, 201
baking soda, 125
banding, 24
bees, 27
Besecker, Thomas, ix, 180
bio-enhancing products, 118
biodynamics, 199
bioelectric interactions, 220
biological activity, 63, 100, 105, 197,
  200, 209
biological farming, 199
biological growers, 146
biological preparations, 201
biological products, 192
biological program, 192, 209
biological recovery, 193
biological system, 208
biologicals, 191, 209
Bird, Christopher, 220
bone meal, 89
boron, 37, 44, 63, 98-99, 167
bottle brush, 166
Bread from Stones, 54

brix, 30, 45, 52, 63, 97, 112, 118-
  119, 128, 145-147, 182-183, 196,
  207; content in trees, 186; reading,
  30
Brunetti, Jerry, ix, 154
BUN, 155
calcitic limestone, 94
calcium, xii, xiv, 2, 13, 22, 35-38, 44,
  46, 48, 56-57, 59-63, 66-67, 69-
  70, 75, 82, 85-86, 88-90, 92, 95-
  97, 98, 101, 166, 172; availability
  of, 93, 146; biologically-active 23;
  chelated, 207; and energy, 93-94;
  liquid, 23, 37, 61, 64, 95, 201-202,
  209; and mineral balance, 111; in
  plant and animal cells, 93; and
  resistance to insect attack, 110;
  shortages in, 182; sources of, 93-94
calcium carbonate, 11-12, 94, 102;
  and pH 105
calcium hydroxide, 102, 105
calcium nitrate, 82, 202
calcium oxide, 95, 105
calcium phosphate, 112
California Department of
  Agriculture, 168
Callahan, Philip, xiii-xiv, 2, 12, 14,
  29-30, 51, 54-55, 133, 179, 215
cancer, 2, 215
capsicum, 31
carbohydrate metabolism, 100
carbohydrates, 117, 150, 192, 196
carbon, 30, 73, 80, 113, 117; active,
  17-18, 54, 192, 208, 142
carbon dioxide, 22, 50, 102
Carson, Rachel, 6
cation, 63, 155
Cation Exchange Capacity (CEC),
  44, 46-48, 50, 54, 73, 87, 123,
  155, 197; soil, 187-188; test, 35,
  37, 59-60, 64, 94, 105, 188; test,
  interpretation of , 43
cellular degradation, 186
cellulose, 146, 160
chelating agents, 95

chemicals, 70, 192; rescue, 188, 194; toxic, 6, 8, 18, 23, 31, 38, 41, 61, 124, 134, 189, 195; and weed control, 176

Chilean nitrate of soda potash, 91

chloride, 45, 67-68, 72, 187

chloride ion, 90

chlorination, 142

chlorine, 37, 68, 90, 98-99, 111, 132

chlorophyll, 79, 96, 100, 160, 179

citrate acid, 99

cobalt, 44, 98-99

Cocannouer, Joseph, 21-22

colloid cations, 106

colloid, 35; clay colloid, 70, 82, 103

colloidal phosphate, 85-86, 88-89, 190, 198, 210

comfrey, 166

compaction, 48, 68, 73, 89, 95, 170-171, 189, 190, 192; caused by equipment, 170

compost, 14, 16, 45, 50, 62, 201, 208

composting, 199

conductivity meter, 132

consultants, 211-212

copper, 37, 44, 92, 98-99; fungicidal qualities of, 100; sulfate, 99

corn, 40-41; earworm, 29, 40; rootworm, 22; fermented, 160; smut, 22

cosmic pipes, 199

cover crops, 18

crabgrass, 48

Crop Services International, 221

cropland, recovery of, 132

crops, genetic potential of, 181

cultivation, 24, 175, 178; for weed control, 179

cultivators, 205

dairy cows, 160-161

dairy operations, 160-161

dandelion, 22

DAP, 82, 87

decay, 50

degenerative diseases, 2

Department of Defense, 133

destructive products, discontinuation of, 187

diamagnetic, 162

diamagnetism, 54, 218

diatomaceous earth, 32

digestive enzymes, 157

disease, and nutrient deficiency, 134-135

disk, 173, 176-177

distillation, 142

Divining Hand, The, 220

dolomite, 22, 40, 70, 94-95, 105; avoidance of, 106, 187

dolomite lime, 95-96, 195

Dow Chemical, 5

dowsing, 216, 219-220

Dyna-min, 154-166; detoxifying effects of, 155; paramagnetic qualities of, 162

E. coli, 140, 213

earthworms, 176

eco-agriculture, xiii, 8, 188, 221

eco-farming, xiv

EDTA chelated trace minerals, 99

EEG, 215

Einstein, Albert, 214

EKG, 215

electrolyte, 79, 202

electromagnetic energy, 152, 216

electromagnetic waves, 215

electronic scanner, 2, 38, 40-41, 64, 68, 71, 76, 86, 112, 114, 188, 220

electronic soil tester, 12

energy, 11-13; absorption, 199; cosmic, 13, 163, 199; eloptic 2, 38; elusive, 215; emissions, 216; field, 57; flow, 19; intensities, 40; magnetic, patterns, 38, 191; positive, 215; readings, 38; release, 203; transfer, 96

Enlivened Rock Powders, The, 54

Environmental Protection Agency (EPA), 2, 6, 28, 32, 133, 137, 140-42 enzymes, 100, 192, 196; digestive, 156-158; and house pets, 162; importance of, 156-158; inhibitors, 157, 162; production, 156-157; systems, 96; transfer, 159

epsom salts, 63, 96, 102, 187

equipment setup, 205

ERGS, 48, 50, 52, 60, 62; ERGS meter, 51, 61

erosion, 181

Exploring the Spectrum, 179, 216

fall programming, 189-194, 207-208

fall panicum, 22

farm management, guidelines, 210-211

farm, model 188; record keeping, 188

farming, biological, 61, 193; chemical, 54, 68, 133; hydroponic, 77; organic, 198

feed, contamination of, 150; dairy, 152; housing of, 152; nutrients in, 150; soaking grain, 157, 159-161; value of, 181

Feed The Soil, 83

Fenzau, C.J., xiii

fertility, 18, 21, 31-32, 34, 36, 38, 41, 43, 89, 110, 172; imbalances, 186; recommendations, 59

fertility programs, 38, 70, 90, 94, 97, 112, 114, 189, 201; biological, 203; conventional, 203; non-toxic, 203; standard, 145

fertilizer, 11, 13, 24, 38, 46-47, 49, 61-62, 64, 68, 71, 73, 79, 176; acid, 23, 195, 202; anionic, 202; balancing, 115; blended, 71-72; chemical, 1, 18, 65-66, 69, 70, 77, 98; cold process, 73; combining, 112; contamination of, 76; dry, 72-75; dry soluble, 198, 202, 209-210; effects on plant growth and fruiting, 110; energy attraction by, 115; energy level of, 73; farming applications of, 114; foliar, 123; hot mix, 73, 196-197; liquid, 13, 73-74, 88; obtaining samples of, 76; preplant, 177, 201-202; producing growth

and fruiting responses, 110-111; quality, 23, 74; radioactive material in, 76; salt, 5, 14, 18, 23, 69, 74; selection of, 78; soluble, 24, 52, 69; solubility problems, 75; sour, 111; stabilization of, 77; standard, 14, 18; starter, 177; sweet, 110; trace minerals in, 99; water soluble, 77

Fertilizers and Soil Amendments, 92

field bindweed, 22

field, biological, evaluation of, 181-186; recovery of, 187-194

field cultivator, 174, 177

field meters, 51

field sprayer, 206

fish, 98, 207; emulsion 201, 207; liquid, 84, 198, 201

fish hydrolasates, 202

flocculate 192

flocculation, 171, 190

fluorine, 98

foliar, 32

foliar feeding, 114, 123-124, 205-206; determining effectiveness of, 126; of forage crops, 147; how to, 124, ingredients, 207; as rescue procedure, 124; testing procedure, 127-129; timing of, 125-126;

foliar fertilizers, fruit producing, 115

foliar spray, 24, 118, 125, 204; for growth, 125; pH of, 125

forage, 145-147; high-brix, 147; quality of, 145-147; quick storage of, 147

formaldehyde, 18, 50, 67, 68

foxtail, 22, 48, 182

free-choicing, 151-152, 163-166

From the Soil Up, 174

fruit trees, 114

fruiting cycles, 109

fruiting energies, 111

fruiting response, 114

full-spectrum lighting, 153

fulvic acid, 208

fumigants, 17

fungal organisms, 163

fungi, 16

fungicide, 5, 182
garlic, 31
genetic potential, 184
gentian violet, 150
ginseng, 199
glauconite, 92, 97
glucoheptanates, 95
grain, soaking of, 157, 159-161
granite dust, 14, 97
green manures, 18, 84, 193; bacterial
    stimulation by, 84; as soil stimu-
    lant, 84; types of, 193
greensand, 92
Grotz, Walter, 143
ground water contamination, 69,
    138-143; cleanup, 141
growth energies, 111
growth cycles, 109
gypsum, 63, 87, 95, 97, 190
Hahnemann, Samuel, 215
Hammaker, John, 193
hardpan, 172, 174-175
harrows, 179
hazardous waste sites, 138
heavy metal, 73, 84, 87-89, 100
hemicellulose, 146, 160
Hensel, Julius, 54
herbicide, 5, 23, 41, 68, 70, 150,
    176, 182
Hieronymus, Galen, 2, 38
high-calcium lime, 11-12, 23, 61, 78,
    94, 106-107, 171, 190
hogs, 164-166
hollow heart, 99, 186
hollow stems, 185
Holographic Universe, The, 214-215
homeopathic remedies, 215
hormones, 16, 149
horsetail, 166-167
humates, 193, 197-198, 208
humic acid, 40, 61, 99, 196, 198,
    208-209

humus, xiv, 7, 16-18,30, 38, 43-45,
    47, 49-50, 53, 61, 66, 68-70, 79-
    80, 82, 86, 92, 113, 115, 170, 173,
    175, 191-192, 205, 208-209; for-
    mation of, 176
hydrated lime, 95
hydrocarbon residues, 189
hydrochloric acid, 155
hydrogen, 44, 117
hydrogen ions, 101
hydrogen peroxide, 126, 142-143,
    145-146, 193
hydroxide ions, 101
hydroxides, 95
immune system, 166
infrared radiation, 2
infrared signals, 29-30, 218
infrared spectrum, 216
insect, 27, 29-30, 32, 38, 41, 66-67,
    70, 132; antenna, 29; attack by,
    113; damage to trees by, 185;
    predator, 31
insecticide, 28, 32, 54
Integrated Pest Management (IPM),
    28, 31
iodine, 98
ion beds, 142
ion exchange, 142
ions, 37
iron, 44, 52, 88, 98, 100, 103, 164
iron sulfate, 99
Johnson grass, 182
Juhan, Deane, 9
kelp, 97
kiln dust, 92
kinesiology, 216-218
Kinsey, Neal, 48, 61
Kuhn, Thomas, 214
lacewings, 31
lactic acid, 160
Lactobacillus, 163
ladybugs, 28, 31
lambsquarters, 21-22, 182
Lamotte, 85, 87
LaMotte Chemical Company, 36

LaMotte test, 36-37, 48-50, 53-54, 60-61, 64, 94, 114, 188, 204; interpretation of, 49
landfills, 138
langbeinite, 92
Lasso, 140
lead, 89
legumes, 184; beneficial, 84; as cover crops, 84; as green manure, 83; nitrogen fixing abilities, 83-84
ley lines, 218-219
lighting, role in animal health, 152-153
lignin, 146
lime, 38, 46, 63, 74, 79, 93, 101, 103, 177, 189, 190, 207; deposits of, 94; lime, 96; and magnesium levels, 106; prill, 188; sources of, 106
limestone, 65, 102, 150; dolomitic, 78
lipase, 156
liquid crystals, 199
LISA, 7
Lisle, Harvey, 54
livestock, and enzymes, 164; immune systems of, 150; nutrition, 149-168; production practices, 149; tips for health of, 153-154; water supply to, 149-150
loams, 35
lodestone, 219
Love Canal, 6
macronutrients, 159
mad cow disease, 213
magnesium, xiv, 22, 35-37, 44-46, 59, 61, 63, 66, 70, 75, 88, 93, 96, 98, 100, 105, 172; chelates, 96; correction of levels, 187; forage problems related to, 106; and pH, 105; sources of, 96
magnesium limestone, 95
magnesium oxide, 99
magnesium sulfate, 96
magnetic energy, 18
magnetic flow, 204
magnetic attraction, 204

magnetic phenomena, 152
magnetic flow, 18
magnetic field, 218
magnetics, 218
magnetism, 218
magnometer, 219
manganese, 19, 37, 44, 98, 100, 166; depressed levels of, 100; paramagnetic properties of, 103; toxicity of, 103
manganese phosphate, 112
manure, 17-18, 50, 62, 82-83, 140, 192-193; bacteria in, 140; breakdown of, 140; chicken, 198; contaminant levels of, 83; as contamination source, 92; energy readings of, 83; and groundwater contamination, 140; improper handling of, 83; and increased cation exchange capacity ratings, 83; liquid, 193; overuse, 92; quality of, 158; ruminant, 199; smell of, 158; as source of potash, 92
MAP, 82, 87, 89
marl, 95, 207
maximum genetic potential, 131
McCaman, Jay, ix, 22, 40, 48, 178, 180
McLeod, Edwin, 83
medicated feeds, 139
metabolic enzymes, 162
Michigan Farmer, 175
micorrhyzia, 87
microbes, 15, 17, 31, 45, 56, 63, 67-68, 79, 88, 98, 191, 209
microbial, 31, 62
microbial action, 60, 62
microbial activity, 24, 48, 57, 59, 64, 69, 75, 98, 208
microbial life, 7, 18, 46, 48-50, 60-61, 64, 66-68, 86, 139 169, 189, 190-192
microbial populations, 90
microbial products, 192
microflora, 166
micronutrient, 98, 100, 160, 163, 166

microorganisms, 15-16, 69, 92, 159, 163, 175, 193
microwaves, 152
mineral, 69; content in food, 133; supplements, 97, 119
mineralization, 190-191, 200
mineralizing, 207
minerals, balance of, 131; deficiencies in, 131; in feed, 151; and insect resistance, 98; and soil health, 131; sweet, 110
mining assay, 37
molasses, 23-24, 63-64, 80-81, 145, 147, 193, 201, 208-209
mold, 150
molybdenum, 44, 98, 100; and nitrogen fixing, 100; toxic levels of, 100
monopole, 14
Monsanto, 5
montmorillonite clay, 168
moon, cycles of, 13, 109; planting by, 110, 114
mulch, plastic, 29
muriate of potash, 67-68, 90; avoidance of, 187
mustard, 23
NASA, 154
National Geographic, 15
National Academy of Sciences, 6
natural meat, 164
nature, laws of, 132
NDF test, 146
nematodes, 16
New Farm, The, 41, 170
nickel, 98
nightshade, 22, 182
nitrate, 98
nitrate nitrogen, 37, 44, 79, 155; excessive levels of, 185
nitrates, 1, 62, 68, 90, 137, 140, 181
nitric acid, 78

nitrogen, xiv, 30, 38, 40-41, 43-44, 49-50, 52, 62, 66-67, 71-72, 78-82, 95-96, 106, 111, 192, 196, 198, 200; ammoniacal, 37, 79, 207; and cows, 155; fixing, 78; imbalance, 111; lack of, 79; liquid, 80-81, 196; and microbial life, 111; microbial production of, 79; non-protein, 161; sources of, 80
nitrogen metabolism, 100
nitrogen oxide, 78
nitrogen release, 45
no till, 33, 175-176
nodulation, 78
nodule, observation of, 184
non-toxic farming, 8, 61, 195-200, 221-222
non-toxic programs, 93
NPK, 34, 66-67, 71-72, 213
nutgrass, 22
nutrient ratios, 51
nutrient ions, 201
nutrient stripping, 189
nutrient uptake, 204
nutrients, biologically active, 86; chemically active, 86; human, 132; role of, 131; soluble, 18; translocation of, 123; water soluble, 37
nutritional sprays, 146
Ohio Cooperative Extension Services, 170
OM reading, 44-45, 60
organic farming, 6
organic movement, 133
organic matter, 16-18, 41, 44, 48, 68, 78-80, 83, 95, 170, 176
ORP, 48, 50
ORP meter, 51, 53
Ott, John, 153
oxide, 99
oxygen, 50, 117, 166
ozone, 142
pancreas, 156
paramagnetic, 55, 163, 194
paramagnetic nutrients, 163
paramagnetism, 14, 51, 54, 218

paramagnetism meter, 51, 54
penetrometer, 175
perennial crops, 204
perennial berries, 114
perennials, 207
pesticides, 5, 23, 31, 41, 45, 68, 70, 137, 150, 182; toxic, 181; water contamination by, 138
pH, 18-19, 22, 24, 44, 46-48, 50, 53, 62, 70, 73-75, 79, 82, 85-86, 89, 93-94, 98-99, 101, 145, 151, 154-155, 190, 197; of animals, 164; to determine calcium need, 105; explanation of, 101-102; manipulation of, 102; and nutrient

availability, 101; in relation to lime, 107; as resistance, 103; and soil type, 105; stability of, 103-105
pH meter, 51, 103
pheromones, 28, 31
phosphate, 13-14, 37, 44, 46, 62, 64, 66, 71, 75, 85, 89-90, 206; colloidal, 14, 191, 197; and energy, 85-86; hard rock, 86, 88, 191; role in weed control, 85; soft rock, 85-86, 88, 97, 190; sources of, 86
phosphate fertilizers, 85
phosphoric acid, 71, 73, 88
phosphorus, xii, 2, 38, 43-44, 46, 66, 72, 84, 87, 95, 112; availability of, 146; biologically active, 88; cycle in plant, 85; feed grade, 150; role in photosynthesis, 84
photosynthesis, 96
pigweed, 21
Pike Agri-Lab,222
plant enzymes, 99
plant extracts, 31
plant fertility, 30
plant evaluation, 182
plants, consciousness of, 215
plow, chisel, 172-173, 178, 205; moldboard, 173-174, 179; spring-tooth chisel, 179
plowpan, 172
pollination, 27

pollution, 138
polyphosphate, 87; energy level of, 87
polysaccharides, 146, 192, 208-209
Popp, Fritz Albert, 216
positive ions, 93, 105
potash, 35, 37, 44, 46, 62, 66-67, 69, 71-72, 82, 89-90, 92, 100, 193, 208; over application of, 170; sources of, 91; sulfate of, 196
potassium, 36, 44, 46, 59, 62, 64, 67, 70, 89, 92-93, 103, 105, 207; biological activity of, 89; excesses, 170; soluble, 89
potassium chloride, 69, 72-73; avoidance of, 91
potassium hydroxide, 73
potassium nitrate, 82
potassium salt, 72
potassium sulfate, 72, 91; energy levels of, 91; sources of, 91
potassium thiosulfate, 97
Price-Pottinger Nutrition Foundation, 157, 159
probiotics, 163
protease, 159, 162
proteins, 191; digestible, 150; in feed, 149
protoplasm, 208
protozoa, 160
pseudo-science, 213-214
quackgrass, 178
quantum physics, 214
quasi-organic, 197
radar, 152
radionics, 199, 216
rations, imbalances in, 150
Reams, Carey A., xiv, 2, 9-11, 13-14, 30, 37, 52, 62, 85-87, 92-93, 101, 105-106, 112, 132-134, 94, 163, 189, 204,
Reams test, 51
redroot pigweed, 22, 23, 182
refractive index, 120-121
refractive values, 119

refractometer, 2, 30, 51, 61, 66-67, 97-98, 117, 126, 132, 183-184, 211; calibration of, 119; how to use, 118; interpreting readings, 119; readings, 57, 85, 118, 125, 128
relative feed value (RFV), 146-147
residue incorporation, 173
residue management, 171, 175
respiration, 96
Rest Of The Story, The, 90
reverse osmosis, 142
Rhizobium, 83
rhizomes, 179
ridge till, 175
rock minerals, 65, 97, 198
rock phosphate, 72, 87
rock powders, 54, 55, 193
roots, 183
rotational grazing, 166
rotenone, 31
rototiller, 177
saddle tanks, 205
Safe Drinking Water Act Amendment, 140
Sagan, Carl, 13
salt, 67, 151, 196; insoluble, 75; residues, 189
salt index, 69
saltpeter, 82
sandburs, 23
scanner test, 55-57
Schriefer, Donald, 174
seaweed, 97-99, 201, 207
Secret Life Of Your Cells, The, 215
seed, 109
seed germination, 190
selective elevator, 196
selenium, 98
septic systems, 139
shade grass, 166
shovel placement, 178
silage, 149
Silent Spring, 6
silica, 167
silver, 98

slag, 207
sludge, 139
sodium, xiv, 33, 36, 44, 46, 50-51, 91, 93, 98, 105,
sodium bicarbonate, 151, 155
sodium meter, 51-52
soils, 41; acids in, 103; bacteria, 84, balance, 30; 95; biologically active, 77, 86, 196; biologically healthy, 67; clay, 35, 47, 70; colloid, 80, 103; compaction of, 70, 170; erosion of, 68; feeding of, 77; fertility of, 24, 100 115, 122, 124, 154, 197; fertilizing, 118; flush, 189-191, 209; high-CEC, 105; holding capacity of, 35; ideal, 169; imbalances in, 70; loamy, 47; low-CEC, 59; microbes in, 44, 72, 84, 111, 113, 115, 189; microbial action in, 86; mineralization of, 191, 204; moisture, 70; muck, 47; nutrient levels of, 200; organic matter in, 79; peat, 35, 47; probe, 33; sample, 33; sandy, 35, 47, 90; test, 24, 33-35, 41, 47, 51, 57; test, interpretation of, 41 43; tilth, 169-172; toxins in, 192
Soul of the Ghost Moth, The, 29
sound generation, 199
sour grasses, 182
soybean meal, 150
soybeans, 157
spider cultivator, 178
split pit, 186
sprayers, 124
spring programming, 201-203
Steiner, Rudolf, 163
stomata, 124
stone powders, 54
Stone, Robert B., 215
stray voltage, 216
Structure of Scientific Revolution, The, 214
subsoilers, 174
subsoiling, 172
subtle energies, 213-218
subtle energy detector, 219

sul-po-mag, 72-72, 91, 97, 208
sulfate, 13, 44, 75, 96, 98; of potash, 91
sulfur, xiv, 37, 44, 63, 66, 82, 87, 92, 96, 98, 100, 112, 191, 209; liquid, 37; sources of 97
sulfuric acid, 102
summer programming, 204
sunlight, 165
super weeds, 23
Superfund, 141
superphosphate, 67, 72, 86-87
sustainable farming, 61
Talbot, Michael, 214
tankage, 84
Temik, 140
tillage, 169-180, 201; deep, 172; implements, 171; night, 179; secondary, 176-177; weed control and, 170, 179
tilth, 15, 68, 115, 171; ideal, 169
tobacco, 31
topsoil, 69
torsion weeder, 178
toxic, 31
toxic reaction, 151
toxicities, 56
toxicity in animals, 166
trace minerals, 44-45, 63, 66, 88, 92, 97-98, 151-152, 156, 191, 198, 208; biologically available forms, 99; organic, 198
trace nutrients, 97-98
TransNational Agronomy, ix
tree evaluation, 185
trees, berry, fruit and nut, 206; foliar spraying of, 206
tricalcium phosphate, 67, 85, 202
triple superphosphate, 195; avoidance of, 87, 187; energy release by, 86

twist shovels, 173
ultraviolet bulb, 142
ultraviolet spectrum, 216
urea, 67, 72-73, 81, 161, 167, 197; energy level of, 81; overuse of, 81; as source of nitrogen, 81
USDA, 2, 5-6, 8, 12, 29, 70, 140; Soil Conservation Service of, 68
velvetleaf, 179, 182
vitality, 40, 55-57, 217; of soil and crops, 133; of weeds and insects, 133
vitamins, 16
von Leibig, Justus, 66
Walters, Charles, xv, 8
water, 137
water filters, 142
water-grass, 22
water pollution, 137
water treatment options, 142
weed control, 178-180
weed pressure, 68
weed seeds, 64, 190; germination of, 179
weeds, 21, 24-25, 30-32, 38, 41, 60, 64, 132; benefits of, 22; growth patterns of, 22
Weeds and Why They Grow, 22, 48
Weeds, Guardians of the Soil, 21
Weeds!!! Why?, 22, 40
Wellhead Protection Program, 140
wetting agents, 189
Wheeler, Philip A., 14, 22, 52, 73
White, Robert, 177
Willis, Harold, 90
Witwer, Silvan ,123
X-rays, 218
zinc, 37, 44, 98, 100
zinc sulfate, 196

About the Authors

Philip A. Wheeler is the technical advisor and consulting agronomist for Crop Services International in Grand Rapids, Michigan. CSI is a soil testing lab and consulting service operated by Phil and his wife Louisa. Prior to his current position, he, along with Ronald Ward and Richard Vaughan founded TransNational AGronomy. In addition to consulting work, he has worked for many years in the areas of research and development of agronomic products and technology. He received a B.S. in science education from Boston University, an M.A. in adult education from Michigan State University, and his Ph.D. in biophysics from Clayton University. He is a national lecturer on biological and sustainable agriculture and its relation to nutrition and health. Prior to his work in agronomy and agriculture, he served in the U.S. Army and then worked as a science teacher in Chester, Vermont. An amateur dowser, graphologist and meta-physician, Phil also enjoys composting and gardening. He is a member of American Mensa.

Ronald Ward grew up in suburban Grand Rapids, Michigan. At the age of nine his parents bought a 50-acre farm 25 miles away from their city home. They spent summers at the farm with

Ron working for neighbors and gaining a love for both the country and for the country life — milking cows, cutting hay and being a young farmer. He obtained a B.S. in park management from Michigan State University; a master's of divinity from Asbury Theological Seminary; and a master's in community counseling from the University of Kentucky. After working for and eventually directing the Lexington Central Kentucky Re-ED Program for emotionally disturbed children, Ron returned to his country roots where he was introduced to alternative health and the Reams method of testing urine and saliva. Subsequently he met Philip Wheeler, who had been trained by Dr. Reams in agriculture, and decided to share the Reams concept on a larger level. It was at that time that he, Phil Wheeler and Richard Vaughan founded TransNational AGronomy. Ron saw a need to provide growers with ready access to the firm's lecture information and together with Phil authored *The Non-Toxic Farming Handbook*. Ron has been married for 30 years to his wife, Dorothy. They have two children — all boys but two — and they live on their eight-acre farm which they call Giving Thanks in the Country.